Petra Opitz, Wolfgang Pfaffenberger (Hrsg.)

Adjustment Processes in Russian Defence Enterprises within the Framework of Conversion and Transition

Beiträge zur Konversionsforschung

herausgegeben von
Prof. Dr. Ulrich Albrecht
Institut für Internationale Politik und Regionalstudien
der Freien Universität Berlin

Band 2

LIT

Petra Opitz, Wolfgang Pfaffenberger (Hrsg.)

Adjustment Processes in Russian Defence Enterprises within the Framework of Conversion and Transition

LIT

Gefördert aus Mitteln der Volkswagen-Stiftung

Die Deutsche Bibliothek – CIP-Einheitsaufnahme

Adjustment Processes in Russian Defence Enterprises within the Framework of Conversion and Transition / Petra Opitz, Wolfgang Pfaffenberger . – Münster ; Hamburg : Lit, 1994
   (Beiträge zur Konversionsforschung ; Bd. 2.)
   ISBN 3-8258-2028-9

NE: GT

© LIT VERLAG   Dieckstr. 73   48145 Münster   Tel. 0251–23 50 91
                              Hallerplatz 5   20146 Hamburg   Tel. 040–44 64 46

Contents

Preface 3

Petra Opitz and Wolfgang Pfaffenberger:
Transformation and Industrial Structure in Russia 5

Petra Opitz:
Newly Emerging Organisational and Institutional Structures During a Process of Adjustment in the Russian Defence Industry 21

Igor Musienko
Management Behaviour in Siberian Defence Enterprises 31

Ella Amosenok and Victor Bazhanov
Specific Features of Conversion in Siberian Defence Enterprises 36

Grigori Tomchin
Conversion and Privatisation: The Experience of St. Petersburg 43

Elena Denezhkina
Problems of Conversion and Privatisation in the Military-Industrial Complex of St. Petersburg 49

Tarja Cronberg
Enterprise Strategies to Cope with Reduced Defence Spending 67

Per Wedlin
A Pskov Electronic Factory in Search of New Customers 84

Mathias Granqvist and Claes Östh
Managing the Conversion of Apparaturi Dalney Svyazi in Pskov 88

Ksenia Gonchar
State Industrial Policy in the Defence Complex 97

Martin Salamon and Ian Whitman
Conversion of the Labour Force: OECD Training Programmes in Russia 104

Peter Lock
Supporting Conversion: A First Approximation to an
Alternative Approach 109

About the authors 120

Preface

Even if the words "weapons export" have actually replaced the word "conversion" in Russian newspaper headlines, the transition of the economy cannot leave the defence sector behind. The defence industrial sector is recognised as one of the basic structures of Russia's industry.

Compared to the 1989 level of military production less than one third of the former military output was left in 1993. This level will be kept up also in 1994. The remaining military production has a special status now with regards to state guarantees for demand and profit. The other two thirds of former military productive resources that are now free from military production need to be oriented towards market conditions.

Many pilot research studies on the micro-economic level in various regions of Russia show a growing differentiation between the manner in which enterprises respond to this challenge. The process of reorganisation of the military sector as a whole and the enterprises in particular is full of contradictions, barriers and the struggle for power by various interest groups.

In the struggle to survive within a changing macro-economic and institutional framework, enterprises were much more oriented towards reorganising their external relations and organisational structures than their internal ones. They built up new networks and associations as a reaction to the very high transaction costs of the high instability of present day market transactions. Even if these do not represent a market solution, they might be a viable interim solution for some period of time. Questions of cost-effectiveness are still put aside. Reduction in output was not followed to the same extent by reduction in employment. Overemployment still exists but a slow migration from the military sector into private business can be observed.

During 1993, privatisation became an important task which dynamised the behaviour of enterprise management. Many managers try to use the privatisation process to add to their rights they unofficially (but de facto) had under conditions of state ownership and extend them to include the right to decide about the allocation of profits as well. The special status of the defence industry and the political influence of the military industrial bureaucracy often meant that the general principles of the privatisation programme were changed into special rules for the defence sector with the intention of excluding almost 30 per cent of the existing military enterprises from privatisation.

At the present time, the privatisation process has only had a small impact on changing enterprise profiles. Real conversion is seldom seen. In most cases, enterprises try to expand their already existing civil production. This is due to the lack of investment and technological gaps as well as the lack of experience in searching for a market demand. Conversion projects are often pushed by existing technological experiences and do not meet a real demand because of the general unstable financial framework (drastically reduced public demand) and the general decline of production (no demand for investment goods). Conversion projects which focus on expanding the production of conventional consumer goods for the Russian internal market are now confronted with growing competition from western products.

Although regional conversion programmes had been worked out for each region in the Russian Federation, they did not help to solve this problem. These programmes lack financial funding except in those regions which managed to get a special share from oil and military exports.

As a policy response to these challenges various concepts of industrial policy are being developed. One of the most influential proposals formulated by defence related interest groups intends to use the defence sector as a locomotive to pull the economy out of the present crisis. Civil reconstruction should be enabled by transferring profits from weapon exports. This concept is led by the fear of losing high-tech potential and the only possibility for integration in the world high-tech market. Financial Industrial Corporations (FICs) should be formed and should serve as key institutions to carry out this concept. However, even if such an approach will help reconstruct and modernise military production, it is completely unclear how it can affect conversion and the modernisation of civil industry. Support for forces of self-organisation within enterprises and strengthening the management potential within them should be the priority.

This volume is a product of our ongoing research on conversion in Russia. It brings together a number of articles analysing conversion from different points of view. Some authors present case studies, some a more theoretical analysis. Some authors live in Russia, some are observers from outside. We hope that this pluralism of views helps the reader to grasp the complexity of the ongoing process. We thank all who helped us in preparing the typescript. A special thank goes to Debra Miles. Last not least we thank the Volkswagen Stiftung for financial support.

Oldenburg, March 1994

Petra Opitz
Wolfgang Pfaffenberger

Petra Opitz and Wolfgang Pfaffenberger

Transformation and Industrial Structure in Russia

1 Introduction

This paper was written within the context of a project which is comprised of an empirical examination of the changes in the organisational patterns of the Russian economy. Special consideration in this paper is given to the armaments industry. The armaments industry has special significance when one looks at the transformation process in Russia. This arises from the fact that the defence sector in the former Soviet economy was extremely important since it was a part of the military system and a part of the centrally controlled economic system. It represents the principle of a command economy in both senses.

Structural deformations in the Russian economy are the result of the centralistic and bureaucratic principle of the Soviet Union's economy. This was characterised by a situation in which political priorities are almost exclusively based on mechanisms which lie outside the field of economics. As a result of development policy and system-specific aspects, the military-industrial sector enjoyed an absolute priority in requesting resources and developing social infrastructure.

In 1988, before the first cuts in military expenditures were made, sixty-two per cent of total production in mechanical engineering was accounted for by armaments production.[1] At the same time, the military-industrial sector does exhibit a certain degree of diversification. Fifty per cent of all non-military consumer durables as well as all civil aircraft, radio-electronic appliances, refrigerators and washing machines were produced in this sector. Eighty per cent of the entire scientific potential in the country worked for the armaments sector. It is possible from these circumstances to derive a material significance of this sector - concentration of modern plants, technological know-how and highly qualified personnel - for the modernisation and restructuring of the whole of the Russian economy and to derive its relevance for transforming the organisational and decision-making structures of the economic system - the economic principle.

---

1 Cf. Jaremenko, Oshegov, Rogovski, Rol konversii v izmeneii struktury ekonomiki, in: Voprosy ekonomiki i konversii, No. 4, 1991, p.86.

It was not possible during the design of the project to fall back on any ready-made transformation theories. Transformation is understood to mean the transition from the previous economic system to a more strongly market-oriented system. Any discussion regarding the transformation is generally characterised by the traditional dichotomy between a competitive economy and a centrally administered economy as it has long prevailed in literature concerning economic systems. According to this dichotomy, in order to achieve a successful transformation, it is necessary to halt the economic process for a split second, create the institutional conditions for a different system, change the awareness of all of the participants in the economic process and then restart the economic process under these new conditions. An artificial interruption in the economic process is just as impossible in the real transformation process as it is in a learning process without any real, time-consuming experiences. All moments of transformation have to take place simultaneously. In this respect, a sudden decentralisation creates a shock to the system which turns the transformation process into a tightrope act between productive and destructive chaos.

A model of gradual transition can be found at the other end of the scale of possible transformation models. It must be shown with this model that more gradual forms of transformation can move in a direction which enhances the sense of individual responsibility and thus permits the development of new co-ordination mechanisms within and between the organisational units of the national economy.

The real transformation process contains elements from the two extremes of the scale outlined in this paper. However, changed forms of co-ordination ("market" instead of "plan") interact with the micro-structures of the national economy. The objective of this paper is to present a few elements of the inner capacity for reform in the industrial structures of Russia. It is not objective of this paper to analyse the current macro-economic situation and its repercussions on the transformation process. A transformation without any grave macro-problems during certain phases is virtually inconceivable. It is important in the long term that independent developments are set into motion from the micro-structures formed by the industrial plants and businesses. This paper, therefore, focuses on micro-structures.

2 Elements of Integration in the Old Economic System and the Need for Reform

Each economic system has a system of rules which possess its own form of rationality. A centrally controlled system has been described as an "economy of shortage"[2] in which individual agents have to try and raise their share on the input side without giving away more in qualitative or quantitative terms to the output side. The lack of efficiency incentives in this co-ordination system has a considerable impact on the development of structures within the economy. A transformation presupposes different co-ordination mechanisms and brings about changes in the micro-structure. This is described in somewhat more detail below.

2.1 Efficiency

Static efficiency exists when improvement in welfare cannot be attained by reallocating production factors. The concept of efficiency presupposes a valuation system which permits a comparison on the basis of products/persons. Although static efficiency is generally considered to be a system-indifferent criterion in economic theory, it cannot be applied to a traditional, centrally planned system in this way. This is because the agents in such a system do not have an appropriate valuation scale and must orientate themselves instead to (incorrect) prices, time-related and location-related availability, and to planned targets. This results in a different form of efficiency logic: efficient persons are those who can satisfy the targets set by the institutions while still observing the bureaucratic rules. At the same time, these persons can also manage to survive within the given set of rules. As a prerequisite, such persons must manage to secure an adequate budget for their activities.

This type of efficiency logic exists in the western economic system as well. It does not apply to the core sector of markets, but is found in many areas of public or quasi-public goods. Vigorous discussion is frequently conducted in the boundary area between the market and the state on how to promote efficiency changes within framework regulations without having to find a clear solution in regards to the specification of demand or the admissible costs.[3] In an economic system with a market-economy core, however, the market

---

2   Kornai, J.. Economics of Shortage. Amsterdam, 1980.

3   Cf. on this subject Scheele, U.. Privatisierung der Infrastruktur. Köln, 1993. (Chapter II is particularly relevant)

economy sector acts as an indirect monitoring and control element since non-market allocation is fundamentally contestable and is thus at least subject to strong potential competition depending on the institutional structure and political conditions.

In centrally controlled economic systems, on the other hand, bureaucratic efficiency was the culturally dominating form. It is hardly conceivable that a far-reaching transformation can take place without a large number of persons who can gain the appropriate experience. This, in turn, calls for the development of a labour market in which efficiency and remuneration have a direct bearing on one another. In the old system, the work-related employment risk was borne by individual enterprises; while in a market economy system it is organised as a fundamentally insurable risk and is, hence, detached from individual enterprises. This is probably where the most difficult problem of the transformation process is to be found: at the present time (irrespective of efficiency and the order situation), employees are not dismissed for social reasons since there are no institutions such as unemployment insurance, instruments of labour market policy, etc., to absorb the impacts of unemployment. Efficiency-specific experience is only possible for the large number of employed staff if it starts at their own workplace.

Up to now, the tasks of the enterprises comprised the production of goods and the assurance of social integration for their employees. A typical feature of western market economies is the division of labour between the market and the state in which social services are not connected to an enterprise. Some large organisations, however, offer a variety of supplementary benefits in addition to wages and salaries - health and leisure facilities, housing, etc. These benefits in a western undertaking mainly represent a part of efficiency-related wage structures. Such benefits serve to improve the working atmosphere and to reduce personnel fluctuation.

Companies are able to exercise this function because the labour market is fundamentally deregulated and the western economic integration inputs - unemployment insurance, health insurance, old age pensions - are not primarily processed via an enterprise.

In Russia, on the other hand, the social network outside the firm is much more loosely knit. Unemployment means the loss of a job and (to a large extent) the loss of membership in a social group. The great importance of belonging to a firm as a basic condition for social integration is a traditional economic attitude in the Russian economy which must be eliminated during the course of the transformation process if the process is to be successful. This necessitates an acceptance on the part of a large portion of the population. Accordingly, the creation of a social security network is just as important for the success of the transformation as is the development of a class of entrepreneurs who exhibit individual responsibility.

The essential prerequisite for more efficient management of the economy is a closer connection between work input and remuneration (even to the point of extending as far as the dismissal of employees). The development of a social security system to cover the previously unknown risk of unemployment is an essential condition for market-oriented development.

Efficiency-oriented pay presupposes an in-house measuring and evaluation system in which individual remuneration can be oriented. Greater differentiation in wage structures would represent a fundamental reform. Firms have admittedly started to adjust pay at the management level by direct and indirect salary elements. This has been accomplished within the framework of greater autonomy which is now enjoyed by the firms. However, development in general (and especially within the armaments industry) proceeds in an opposite direction from efficiency-oriented differentiation. In a situation characterised by declining orders, direct and indirect wage payments are essentially a measure of preservation and social stabilisation.

2.2 Dynamic Efficiency

Dynamic efficiency relates to the adaptability and innovative capability of an economic system. The transformation process is burdened by structural disproportion which has evolved as a result of central control and political disagreements. Dynamic efficiency calls in particular for de-centralised areas of responsibility and presupposes a variety of competing blueprints for the future. Parts of the armaments sector in Russia have to switch over their production in view of the large surplus in production volume. They are well equipped with technological resources, highly qualified personnel, etc. In structural terms, therefore, the conditions for mastering conversion and producing new products and processes are favourable. However, empirical investigations show that such processes are rarely set in motion and are frequently not even desired.[4] This is probably due to the unfavourable macroeconomic factors - inflation, shortage of capital, etc. - to the lack of experience in dealing with uncertainty and to the lack of decentralized responsibility. A firm's management system is not geared towards making changes in product development based on the management's own decisions. The

---

4 Cf. Opitz, P. (1980). <u>Das Dilemma der Rüstungskonversion in Russland</u>, in: Osteuropa-Wirtschaft; Vol. 2, 1993.

problem of dynamic efficiency at the level of a firm is essentially a problem of redeveloping activities starting from the top management level. The skills needed by a market-oriented manager are different to those required by a successful administrator in a planned economy system (see above).

2.3 Vertical Integration

The advantages of interlinking between production stages and the associated reduction of uncertainty play a major role in all economic systems. In the western economic system, market strategy criteria such as, securing sales and distribution channels, monitoring/controlling procurement sources for competitors, etc., are of great importance. Such aspects barely played a role at all in the "economy of shortage". Many firms believed that the objective was to reduce uncertainty, thereby taking economic inefficiency into account. Vertical integration is a moment of demarcation between one's own organisation, its successful survival and the outside world. The high degree of integration and the reduced importance of external relations was (as seen by the authorities) a prerequisite for controlling and monitoring these enterprises.

Vertical integration is only rational if the most cost-effective scale can be reached during all of the integrated production stages. If this is not the case, participation of the direct vertical integration would be organised instead, as in the classical approach in the western world. It would be possible in this way to secure the services of a production stage by means of market strategy activities and to secure the surplus capacities of those parts which are not fully utilised by the organisation's own demand - this surplus could be sold on the market.

The degree of vertical integration in the traditional Russian economy cannot be measured with regards to its economic efficiency since empirical documentation does not exist. In the "economy of shortage", vertical integration was the consequence of a survival-oriented attempt on the part of the firms to secure self-sufficiency. It can therefore be assumed that the degree of vertical integration extended well beyond the optimum level. There is potential within the context of the transformation process for improving efficiency by doing away with rigid vertical integration and by replacing it with less strict forms. This generally presupposes a break-down in the enterprise's present complex structures. If this does not occur, the parts of the enterprise which show a deficit will be a burden on the development of the whole enterprise. If it does occur however, surplus capacities will become apparent which were not previously evident.

In the real situation of the present Russian economy, such rational treatment of this train of thought is hardly practicable since the internal information systems in the firms are not developed enough to derive appropriate conclusions. First of all, it is necessary to develop a management information system that can serve as a basis for corresponding management decisions. Previous accounting systems were designed for quite different purposes.

The development of an operating information system which can supply the management with a basis for a corporate policy oriented towards efficiency and the market is of central significance. Such steps calls for the development of appropriate guidelines, personnel training (so they can perform the work) and management personnel who know how to use these instruments.

2.4 Economies of Scale

Economies of scale are achieved by enlarging the production scale and are typical for industries in the growth phase. Growing markets allow a reduction in cost. As previously mentioned, it is likely that the potential will not be fully exhausted in view of the excessive vertical integration. On the other hand, wrong decisions have also been taken in the opposite direction, presumably on the grounds of political directives. A consideration of this aspect must, after all, cover the complete product chain including transport. Incorrect presentation of transport costs in the specified prices for transport services can, like the politically motivated target directive of regional division of labour, lead to excessive concentration at a few locations.

In this case, a reform from the inside is barely conceivable. The most cost-effective measure of production and spatial division of labour can only be found by the development of competition and by the establishment of new firms. At the present time, there is a strong inclination to enter into inter-firm co-operation. Nearly all the surveyed firms reported activities within the framework of associations. Such associations probably perform the type of crisis management needed at the present time and represent a replacement for the loss of central co-ordination. During the course of time, such structures could prove to be a considerable obstacle to competition.

Similar structures existed in Great Britain and France after the Second World War.[5] Modernisation commissions (France) and working parties

---

5   Cf. on this point e.g. Denton, G. et al.. <u>Economic Planning and Policies in Britain, France and Germany</u>. London, 1968; Shonfield, A..

(Great Britain) consisted of representatives from the branches of the economy and the government. They were intended to function as the "transmission belt" of an innovations-oriented economic policy. Although such structures are not acceptable from the regulative point of view of a fully-fledged market economy,[6] they can, however, represent an interesting instrument if it is possible to reorient them away from a defensive into an innovative position within the context of a transformation process. Lobbies or pressure groups, as are quite commonly found in a fully-fledged market economy, do not yet exist. Such associations represent the beginnings of this type of process.

2.5 Economies of Scope

Economies of scope are an important indication as to whether the structure of a pure market economy tends to develop in the direction of an organised market economy. It is difficult to interpret the horizontal links of Russian enterprises in a manner which is comparable to the usual links in a market economy. Infrastructure inputs - housing, social services, road building and road maintenance - are frequently produced at the same time as industrial products. Such interlinkage in production can be rational in a certain development phase to minimise transaction costs. During the development phase such interlinkage ensures that infrastructure services and production are commensurate with each other; a planning bureaucracy would presumably fail to co-ordinate the production of such complementary goods.[7]

This logic later gets lost in the fog of historical development. Many links in the production system today are simply the product of political intervention. Testing such links on economic terms can be an important momentum of reform.

Synergetic effects between different lines of production are possible and likely. These effects are generally difficult to evaluate; synergetic flows be-

---

Modern Capitalism. London, 1969; Lutz, V.. Central Planning for the Market Economy. London, 1969.

[6] This was emphasised, in particular, by the German discussion on French planning. Cf. for instance Hedtkamp, G.. Planification in Frankreich. 1966.

[7] In Williamson, O.E. Die ökonomischen Institutionen des Kapitalismus. Tübingen, 1990. (The construction of works housing is given as a classic example; cf. p. 40 ff.)

tween parts of an enterprise often develop into cross-subsidies. Parts of an enterprise must then support the other parts.

It is hardly conceivable that trade flows between the various branches are balanced in any major organisation. As in the case of vertical integration, a snapshot analysis of the advantages of interaction will probably refer to historical fortuitousness in its explanation of co-operative links. Western structures are specifically characterised by a broad range of combined actions which are offered by the legal framework and by information systems which allow an economic evaluation.

By analogy with vertical integration, the moment of political intervention and the possibility of monitoring and control by political instances play a major role. This explains why it is necessary to increase the prospects for new demarcation and co-operation.

As set out above, vertical and horizontal concentration extends well beyond the measures which can be reasonably justified in economic terms. This structure considerably inhibits the processes of adaptation. It would be expedient to divide up enterprises in order to approximate the sizes and the terms of reference that a potential market needs within a framework of privatisation. In place of self-contained integration within the framework of a conglomerate firm or undertaking, more flexible structures could be introduced so as to allow the compound system (still technically necessary) to exist without cementing the old, tight unity and the old command structures.

2.6 The Role of Infrastructure

The division of labour between government and enterprises is organised differently in a market economy system than in the case of the former Russian system. Although the demarcation between infrastructure and normal markets produces a few headaches for the western economist (especially in the boundary area between the two systems), Russia has such a large number of clearly defined infrastructure facilities, which are operated by the production undertakings, that there is no need to discuss the difficult demarcation problems here. It is true that there would not be a problem setting out the organisational and technical infrastructure departments of firms differently in future. The main problem, however, consists in the maintenance of such departments. These facilities have so far been co-funded by cross-subsidies which originate from the income flow of the undertaking; they have not been financed by public sector budget funds. Thus, firm-specific and location-specific "infrastructure duties" are applied instead of general taxes. This situation cannot continue in the long term; the distortions of competition are too large.

3 The State of the Enterprises

3.1 Current General Conditions for Enterprises

The shock therapy that was introduced with the deregulation of prices at the beginning of 1992 was unsuccessful as an attempt to stabilise the national budget. The lack of success resulted from the transfer of a monetary, macro-economic concept to the Russian economy without taking into account its own specific characteristics. These characteristics primarily include:

- a high level of monopolisation in production;
- a lack of institutions (including a well functioning money economy) that conform to market conditions;
- a lack of flexibility which resulted from the old economic system and prevented a swift modification of production functions;
- and the forced severance of previous division-of-labour relations between enterprises and regions caused by the disintegration of the Soviet Union.

The policy applied under these conditions led to a drastic decline in industrial production by almost one third compared with the preceding year, a high inflation rate (approximately 1500 per cent) and an increase in the previous structural disproportion.

A change from the former economy of shortages to an economics governed by sales crises took place. In the first half of 1992, a high rate of inflation coupled with a restrictive budget policy caused the solvent demand to shrink and altered its structure. The insolvency of individual enterprises and the poor payment discipline of even government orders led to a chain reaction which produced dramatic inter-firm indebtedness that reached a volume of approximately 3,000 billion roubles by July 1992.[8]

The previous centralistically organised allocation and distribution system was cancelled and enterprises were left to manage on their own. The infor-

---

[8] Sapir, J. <u>Inflation, Depression and Stabilisation in Russia: Why Traditional Macro-Economics Have Failed,</u> Working Paper: Ecole des Hautes Etudes en Sciences Sociales. Paris, 1992; p. 14.

mal structures, which, in the old system, used to function alongside the formal structures, filled the vacuum left by this cancellation.

These conditions applied to the whole of the industrial field (apart from the basic material industries) and has had an even keener impact on the military-industrial sector. Cutbacks in orders amounting on average to 68 per cent in comparison with 1991 (in individual areas cutbacks even amounted to 90 per cent and extended as far as the cancellation of individual projects) as well as shrinkage and structural changes in the non-military market made it impossible for enterprises to achieve a balance on the basis of their own potential.

In order to avoid a further decline in production and, in the final analysis, a disintegration of the production potential, the military-industrial sector was subsidised in 1992 by payments from the national budget which altogether amounted to 623 billion roubles. This was done on the principle of equal shares for all.[9]

The reform in institutional terms led to the abolition of the old centralist control system. The Ministries of Industry, which in addition to issuing orders also guaranteed supplies of appropriate materials and investment funds for the enterprises, were dissolved and replaced by branch-specific Departments. The Ministry of Industry was left with a much reduced staff. However, these Departments were unable to perform their original distribution functions. A government committee for the defence branch was set up during a further reform phase for the armaments sector. This committee is responsible for issuing orders for military procurement and is one of six different authorities for conversion.

The withdrawal of state control for enterprises was supplemented, at the same time, by an attempt at forced privatisation for state-owned enterprises. This attempt initially comprises commercialisation of the industrial enterprises by creating joint-stock companies; it does not constitute the real rights and duties of ownership.

---

[9] 77 billion roubles low-interest credits at three per cent per annum., 46 billion roubles direct grants (both items, however, were used primarily for the payment of wages and salaries) and at least 500 billion roubles to balance inter-firm indebtedness. Cf. A. Lomanov. Oboronka ne vyderzala ispytanie rynkom, in: Inzenerskaja gazeta, No. 6/1993, p. 2.

3.2 The Behaviour of the Enterprises

3.2.1 Internal Changes

The Gaidar government took the economic situation into account because of the radical cutbacks in orders for the armaments sector (necessary to restore the budget to financial soundness by reducing public expenditure). The government also intended to exercise pressure in order to achieve a reduction in armament-oriented capacities and to reduce the military-industrial structures of power. However, the comprehensive and uncontrolled demilitarisation of an economy, in which the essential industrial cores are at the same time characterised by these structures, entails the risk of partial de-industrialisation and hence of social collapse.

The structures of the military-industrial sector that cannot be periodically characterised as anti-reform - the victory of Yeltsin and the democrats was possible only with the consent and/or toleration of these structures - reacted to the changes in the outer conditions by orienting themselves towards their direct interests of survival and the retention of their potential.

The observable changes at the micro-economic level were comparatively small when measured against the considerable changes in the external conditions. A low level of fluctuation did not adequately reflect the actual underemployment situation.[10] Explicit dismissals have only taken place to a minor extent; pensioners were the main persons to be affected. In addition, there is a high, voluntary fluctuation rate which results from the low wage level in the public sector and from the level of employment in the defence sector in comparison with the private sector.

Current income levels are maintained by primarily redistributing funds earned from existing orders and by distributing funds from government subsidies for conversion. This is done virtually independently of an employee's real share in the work.

Only the first tentative approaches towards de-concentration and new combinations of internal structures can be observed at the present time. Differ-

---

10   In 1990, 300,000 employees in the defence sector were made redundant; however, 228,000 were re-employed in the same year. In 1991, the ratio was 380,000 to 300,000. Source: Y. Kuznetsov, A. Ozhegov. <u>Transformation of the Russian Defence Sector: Enterprise's Behaviour in 1992 and Prospects for the Future</u>, manuscript.

ences also exist between scientific institutes (NII) - most are only composed of an affiliated model establishment - and production enterprises. Repair and preservation services in the institutes have been removed and set up as independent small-scale firms which lease or rent industrial buildings and equipment. The rationale for proceeding in this manner was derived from the relatively low degree of an institute's utilisation capacity; a part of the employment risk was shifted in the manner to the outside world. At the same time, splitting-off agreements generally ensure that contracts from the parent organisation retain a priority. New fields such as marketing are also established in some cases outside an existing enterprise. These partial changes are not yet in line with a fundamental internal re-organisation and a re-organisation of decision-making structures within an enterprise. The conversion of large-scale undertakings into holding companies is generally of only a formal nature; but, it potentially renders structures more effective by developing the means of independent economic responsibility for each of the individual member firms.

3.2.2 Changes in Inter-firm Relations

A large number of enterprises are forming horizontally oriented groups and associations in order to compensate for the loss of previous institutions of vertical co-ordination and the lack of macro-economic control. There are regional differences in the activities which lead to the formation of such branch-oriented structures and their organisation. These institutions represent a kind of self-help mechanism, which is, however, generally based on a non-economic relation and a non-monetary exchange.

Agreements are made on the basis of former informal relations in order to avoid competition and ties to a potential customer. "Pocket banks" established by the enterprises only grant preferential conditions for their shareholders. These banks function with other commercial banks as a channel for the distribution of subsidies issued via the Central Bank. Enterprises can in this manner bypass market requirements or make use of the scope for financial transactions.

4 Options for Economic Policy

The current economic policy considerations of the Russian government are focused on maintaining and stabilising existing large-scale enterprises because the experiences of last year - the attempt at shock-type deregulation of prices and the collapse of former relations between enterprises - help to critically determine the severe recession.

The State's withdrawal from economic control of industry, which was deliberately quickened by the Gaidar government, was aimed at the necessary breaking up of the old structure. Such steps were necessary for psychological and political reasons even if it was controversial from a point of view of economic theory. The broad consent to continue the process of reform (despite criticism) shows that these steps were necessary. However, a sudden and fundamental reduction of government control, while conditions of non-existent compensatory market institutions and major structural disproportion exist, only appears to intensify the crisis. A gradual sequence of economic policy measures, which are sensitive enough to avoid a further decline in production and loss of equipment as a result of long-term use, is necessary. Short-term objectives must be coupled with long-term options. The government is expected to perform a tightrope act between phasing out and converting old structures while depriving them of power without making them totally inoperable. This is a stop-and-go policy within an evolutionary process.

An unconditional de-concentration would not be very helpful for large-scale enterprises which are generally structured along interlinked technological production channels. This would not improve effectiveness. On the contrary, it could only end by paralysing the production process. Some de-concentration can be economically expedient especially when accomplished in connection with actual conversion.

Although conglomerates of enterprises and banks are emerging as a self-help mechanism for large enterprises, they fill the vacuum left by previous vertical control and distribution mechanisms and cannot be clearly characterised in regards to their operating mode and future role. Such structures are effective as a stabilising factor and they provide a possible form for accumulating productive capital.

Public funds are still necessary to replace the orders from former defence contracts. However, the manner in which these funds are awarded should change from the previous form of maintenance subsidies to subsidies which promote adaptation. The appropriation of such funds to aid specific conversion projects would be a step in the right direction.

The privatisation of state-owned enterprises is necessary in order to relocate responsibility and decision-making powers within the level of a firm. The type of privatisation which is currently taking place - privatising large state-owned enterprises by creating joint-stock companies - is really a more formal procedure. It can be merely considered as a de-nationalisation of the undertakings.

Although a lack of transparency in privatisation variants and their economic consequences exists, the privatisation process has encountered great interest. Most undertakings in the armament industry still strive to achieve privatisation despite the contrary intentions of the government Committee for the Defence Sector. Management tries to circumvent the dictates of the various state institutions since this is the only way to acquire the appropriate rights of disposal. (Real financial support can hardly be expected.) Other contrary examples exist and arise from enterprises who can only foresee slim prospects for the future.

It is uncertain whether these first steps towards creating an entrepreneurial system from the former large enterprises will in fact conform with market economy requirements. Growth on the basis of profit-oriented business administration also calls for strategic concepts and independent surveillance systems in the form of supervisory boards. One of the most critical shortfalls in the present privatisation model can be found here. (The financial rulings are another aspect.)

It is too early for legislation to govern competition; however, an institution should be created and be made responsible for observing the development of industrial structures from the point of view of competitive aspects. It should also be committed to the concept of competition. This step could eventually help in preparing legislation which governs competition on the basis of information concerning actual developments.

Conditions of competition under the described conditions can only be gradually created by a variety of activities which work at different levels. It must also be based on the assumption that new businesses will be selectively promoted.

Competition for the dominating enterprise structures can only be created by building up additional, new capacities outside the previous large-scale undertakings. This could be done with foreign financial assistance and with a mix of domestic and foreign capital with capital coming from the private sector and the state-owned sector. Greater decentralisation and regionalisation of economic development could play a positive role in this respect.

The establishment of small-scale firms in the ancillary sector should be promoted alongside the settlement of businesses. Efforts achieved by partitioning components of an enterprise are no longer an exception even in the defence sector. It is necessary for the separated firms to achieve a balance between the creation of conditions for economic and legal independence - infrastructure and markets for products - and contractual obligation to complete work for the parent company. The risk of interruption in the production process by an insufficient number of alternative suppliers of analogous inputs should not occur.

Economic policy measures implemented to develop a stable and diversified banking system and the improvement of a preferential taxation system to promote productive investments are aspects which could promote entrepreneurial systems. It is necessary to establish government-operated or public-sector development banks in order to achieve free access to capital that is independent from informal and non-monetary connections (pocket banks). This is vital for new businesses. The participation of private structures should not be fundamentally excluded. Such structures represent a counterweight to the growing economical and political power of the newly formed conglomerates. It also represents a counterweight to the exchange and financial structures which these institutions have financed and continue to control.

Petra Opitz

Newly Emerging Organisational and Institutional Structures During a Process of Adjustment in the Russian Defence Industry

Russia still seems to paint a very frustrating picture if one focuses on the macro-economic level.[1] But, this is only one side of the picture. An increasing discrepancy between the macro and micro-economic level has existed in Russia during the last two years.

The focus of our project is to change the behaviour of defence enterprises in two different regions: St. Petersburg and Novosibirsk. Our information consists of interviews, conducted by experts, with directors, members of enterprise management, members of a city's executive and legislative institutions and representatives from corporate structures or associations which are concerned with the defence sector. Interviews were carried out in both regions with enterprises from different industrial sectors: machinery, heavy machinery, the aircraft industry, the electronics industry and small gadget manufacturers. We carried out additional interviews with manufacturers of shipbuilding, information and communication systems in St. Petersburg.

The economic and social structure of both cities is highly militarised. Although available figures should be treated with caution, statistics show that between 1989-90[2] 70-75 per cent of the total industrial output in St. Petersburg was concerned with the military-industrial sector while, at the end of 1991[3], 60 per cent of the total industrial output in the Novosibirsk region was concerned with military production.

In 1992, cutbacks in military orders amounted to an average of 60-70 per cent in comparison with 1991. Individual areas have experienced cutbacks amounting to 90 or almost 100 per cent - Baltijslij Zavod in St. Petersburg and Sibselmash in Novosibirsk. These cutbacks occurred in conjunction with a shrinkage in industrial capacity. Structural changes in the non-military

---

1   Die wirtschaftliche Lage Rußlands - Monetäre Orientierungslosigkeit und realwirtschaftlicher Aktionismus, DIW Wochenbericht 42/93.

2   According to unofficial statements made by experts from the St. Petersburg Mayor's board.

3   According to analysis of STRATECON, unpublished manuscript, Novosibirsk, 1993.

market have also made it impossible for enterprises to achieve a balance on the basis of their own potential.

Macro-economic decisions were paralysed by the struggle of different political power groups. as a result, many necessary decisions were not made, for example:

- a military doctrine that sets figures for stability on further military procurement or development expectations was not approved until the beginning of November 1993;
- an industrial policy was not established to seriously include conversion in the industrial-sector restructuring process (conversion is pointed out as a separate direction on the list of prioritised development directions);
- until now, there has not been an identification of firms which should be closed down;
- decisions concerning employment policy have been lacking.

The State, a part of the transition as its former functions are consistently dissolved, lost its interventional power. It is also not even able to set the necessary political, economical and legal frameworks and is not capable of regulating the economic process. (This factor is even more relevant for the defence sector since the State remains the most important consumer for a part of its products.)

As a result, enterprises had to muddle through this mess. Many survival measures were undertaken such as, selling high-quality and deficit materials supplied at low prices to the military sector, searching for any order to use existing production capacities or setting up small-scale firms to lease or rent industrial buildings and equipment. In general, the reduction of output was not followed by a respective reduction in employment. There was only a voluntary fluctuating rate which resulted from the low wage-level that now exists in the defence sector (it is now the lowest wage-level in the state owned sector).

Productivity has also declined while cost efficiency was and is still not achieved. The majority of enterprises could be named hibernators since they try to survive without principally changing their internal structure. Except for a few enterprises, during the first half of 1992, a general tendency towards lethargy and the demand for subsidies from the state continued to dominate. However, in the second half of 1992 and during 1993, a more dynamic behaviour of a large number of enterprises was observed. Increasing activity and differing types of behaviour could be summarised by the concept of self-organisation - it is based on a hidden rationality of economic agents. This process includes two key elements: privatisation and the creation of corporate structures. Both elements are mutually connected. Sometimes they occur simultaneously or sometimes the latter is result of privatisation.

Privatisation started with a political initiative from "above". The initiating role of political institutions is still visible. For example, if you compare the dynamic of privatisation in St. Petersburg and Novosibirsk, St. Petersburg plays the role of an avant-garde. Forty per cent of the 570 enquiries for privatisation received by the city's Committee for Management of State Property (CMSP) came from military enterprises. This characterises a growing interest in the military sector towards privatisation.

By contrast, the privatisation process in Novosibirsk is proceeding more slowly. The pressure for conversion and for the creation of structures capable of withstanding market pressures has been diminished by short-term profits arising from new arms export deals and by continuing subsidies. The regulatory power of state institutions is underdeveloped. The social and civil infrastructure is dependent to a far greater degree on the arms industry. A multitude of corporate structures have emerged here. The structures are more strongly oriented towards the direct regulation of economic activity "from below" and towards what in some cases are cartel-like links between enterprises. The privatisation of defence enterprises has in some cases completely stopped.

Thus the local constellation of political forces and the established contacts of co-operation with western companies are important factors which influence the dynamic of privatisation forces. While in St. Petersburg the CMSP is a major factor of political influence within the constellation of the existing Soviet (Petrosoviet) and the Executive (Mayor's board) - in Novosibirsk the Soviet and Executive both are dominated by members of the old party nomenclatura.

Although the present legal basis for privatising defence enterprises gives different interest groups plenty of room to quarrel and although Presidential decree Number 1267 (drafted in August 1993) seems to create difficulties for some 474 key enterprises, privatisation allowed management the opportunity to obtain legal ownership and control of their enterprises. Military production at present is economically unattractive for many enterprises. There are already examples of former defence enterprises rejecting military orders from the state. The necessary investment for civil production is expected from western customers or co-operating partners, who hardly want to give money to state owned military enterprises.

Even if the actual privatisation process establishes new governance structures for the enterprises and gives an impetus towards:

- commercialising enterprise activities,
- changing an enterprise's organisational and physical structures,
- improving the operational discipline of the management,
- and introducing intra-organisational cost accounting,

the most important impetus is the expected opportunity to attract foreign or national loan capital and expertise. (A special method of including an investor's interest was worked out by the St. Petersburg's CMSP.)[4]

The very controversial privatisation process carried out by Baltijskij Zavod demonstrates this point since, in this case, the controversy between the State Committee for Privatisation (Chubais) and Roskomoboronprom (Gluchich) was settled after the intervention of Yeltsin himself. The shipbuilding company, rich in traditions, previously in the past produced nuclear ice breakers, battleships, naval equipment, steam producers for the nuclear drives of warships and submarines. Today, the outlook for production of warships is bad and the dependency four times as much as in the field of civil orders. (The cancellation of an order to produce a battleship sets free 50 per cent of the enterprise's capacity.)

After facing the continuous reduction in military orders from 1988 to its almost complete stop in 1992 (6 per cent of military orders remained), the re-entry into merchandise shipbuilding was the only chance for the enterprise to survive. There are good opportunities for the firm in the civil market. Nearly all of the percentage of shipbuilding capacity is used for executing expected orders from Western Europe, especially the Federal Republic of Germany and Norway, for Roll-on Roll-off ships, chemical liquid tankers and others types of commercial ships.

Depending on the selected model of privatisation and the different fractions of shares held by the government or its agencies, an enterprise's management could have more space for decision making or the risks could be higher if the control by a government agency is maintained. An enterprise would be

---

4   see G. Tomchin's paper.

less likely to go bankrupt if this occurred. The government would try, in this case, to rescue the firm in order to avoid the social costs of liquidation.

Some enterprises still do not want to be privatised because they are insecure about their chances for future development. This is often the case in huge machine-building enterprises from the defence industry which currently suffer radical reductions in military orders - Sibselmash in Novosibirsk, for example.

Although there are pioneering examples of separations of old enterprises, de-cartelisation of big firms has not occurred to any large extent. The separation into juridically autonomous economic subjects is very slow and contains backfalls. Many small firms once separated out from big enterprises have been reintegrated. The reasons are, the general difficult economic environment, the lack of state support for small and medium-sized enterprises, the differing interests within big plants and any unsolved problems concerning the actual separation of social infrastructure from an enterprise's holdings.

At the present stage of reform, ideas, such as creating holding companies, seem to be very attractive for a lot of big military enterprises. This allows production units to separate into independent agents in almost all spheres of economic activity. Joint ventures can then be established with separate units. (This is impossible when a whole enterprise is concerned.) At the same time, an intermediary is set up and is able to work out and co-ordinate development strategies, to co-ordinate distribution and to cross-subsidise units. (An intermediary can be state-owned or private.)

For example, between 1992-93, the Arsenal company privatised its civil production parts and transformed them into separate joint-stock companies: Arsmas, Arsto, Arkom, Mars and Arton. The whole company, including the military production parts, is now transformed into a holding company. The confrontation of its economically autonomous working joint-stock companies with market conditions led to a partial correction of prior conversion approaches. One of Arton's conversion projects concerning the elaboration of modern medical techniques on the basis of a statistically calculated demand of Russian hospitals was terminated. This change reflects the dramatic situation created by the present economic and social crisis. There was simply no real demand for this type of service because hard budget constraints reduced budget deficits and because the continuing system of no-tariff health care did not require their type of service. Arton now produces gas pistols to bring it out of its deficit balance.

What impact will privatisation have on conversion and structural changes? Actual changes in output structure were due to dramatic reductions; however, it does not mean that restructuring has taken place. The decline in military production (some 50 to 60 per cent) was not compensated by increasing civil output. Civil production also declined in many enterprises. Changes in equipment and labour force structure did not follow changes in output structure. Thus conversion in the expected matter did not take place.

It is difficult during the actual stages of reform and privatisation to trace enterprise development only back to privatisation. Some estimates are available on the expectations and reflections of managers of joint-stock companies in the defence sector. As result of an inquiry[5], it has been stated that:

- managers are more optimistic in evaluating their financial or economic situation;
- managers are more strictly connecting supply and demand;
- the share of enterprises operating on reduced working hours or near bankruptcy is smaller in the privatised sector (41 per cent) than in the remaining state owned sector (51 per cent);
- 17 per cent of joint-stock companies plan to increase employment as compared with only six per cent of state owned enterprises.

At the present time, it is hard to determine the real impact of this data.

In general, the selected model of privatisation reflects the cleverness and interests of enterprise directors and management. At the same time, the process shows an increasing differentiation between enterprises and their management. It actually seems that three different groups can be identified:

1. Those who do not require state subsidies but law, order and stable macroeconomic conditions. These persons behave like rational entrepreneurs; they try to reorganise organisational structures and restructure production - input, output and the relationship with suppliers and consumers. They anticipate competitiveness on the world market, co-operation with foreign companies and opportunities to attract foreign investment. Some of them try to reduce military production and mobilisation potential. They mostly do not intend to create or participate in vertical integrative

---

[5] Enquiry carried out by the Centre ekonomiceskoj konjunktury pri Sovjete Ministrov - Pravitel'stve Rossijskoj Federacii, September, 1993.

structures. This group usually does not include typically large enterprises.

2. Those who have learned how to behave under the new conditions, but require continuing state support. Their activities are generally directed at surviving by extrapolating existing technical and human capacities and by trying to commercialise an enterprise's assets. They do not usually undertake measures to principally reorganise their structure. Privatisation is used here to some extent to separate production units into judicially, autonomous economic subjects under the intermediate governance of structures shaped in the form of holding companies. They require investment and, with a few exceptions, their products are not competitive on the world market. The enterprises belonging to this group often form associations or vertical integrative structures.

3. Those who are unable to diversify or carry out a programme of conversion. Such enterprises would like the old system of state orders to be restored so as to provide them with the necessary materials and investment for their products. Such enterprises could close after a period of time.

The growing differentiation between enterprises is due to such subjective factors as the personality of the director (even if directors always behaved like patrons), the behaviour of the management, the technical structure of the equipment, the enterprise's products as well as the industrial sector in which the enterprise belongs. Each of these factors affects the future chances of an enterprise, the size of the enterprise and their former share of military production.

The creation of corporate structures as an alternative to regulation is another form of activity that is undertaken by defence enterprises. These associative structures form a distinctive fourth institutional basis of order[6] and have become increasingly important since the classical institutions, such as the community, market and state, do not function efficiently.

The St. Petersburg Association of Industrial Enterprises (founded in 1989), the St. Petersburg Association of Privatising and Privatised Companies (founded in April 1993 as a local group of the Association of Privatising and Privatised Companies in Russia which is under the guidance of Gaidar), the

---

[6] Cf. W. Streeck, Ph. C. Schmitter, <u>Community, Market, State and Associations? The Prospective Contribution of Interest Governance to Social Order</u>, in: Markets, Hierarchies and Networks,. London, 1991, p.228.

St. Petersburg Military-Industrial Corporation (founded in 1992), SIKOM (founded 1989) and Sibagromash (founded in Novosibirsk in 1993) are examples of such structures. These organisations are corporate structures of various different types. Their aims differ and have also changed over time; however, they have all been seen as attempts to adapt to the disappearance of the previous administrative, centralised, regulating structures. Although the large majority of defence enterprises continue to be state owned enterprises, institutional responsibility for the enterprises is not yet clear. Previously existing structures, such as the Governmental Committee for the Defence Branches, the Ministry of Defence and the Committee for Structural Policy, cannot fulfil the functions previously provided by the State.

Functions of production, such as co-ordinating the division of labour and organising distribution systems, cannot be handled by the market because its structures do not yet work. Corporate structures, therefore, compensate for this deficiency. Agreements are made, projects are developed and joint statements towards economic-political decisions are formulated on the basis of informal relations between the former industrial elite. Joint banks, investment funds and various marketing societies are also developed on this basis.

Although the associative structures have mostly been developing into commercial operations, it is interesting to note that some of the corporate structures are of different importance and operative direction in the different regions. Some of the associations in Novosibirsk function as a pool or trust whereby functions of regional regulation and regional industrial policy can be implemented. The lack of state regulation influences the situations of the defence enterprises and requires counter-active measures. These types of structures in St. Petersburg have a more political character since they often carry out different reform approaches. The consequences of conversion are less dramatic in St. Petersburg. More co-operation with western companies can be found there and the infrastructure is ample and flexible enough to solve social problems since there is employment in the new private sector and alternative socially integrating structures. The city is also less dependent on defence enterprise infrastructure.

These institutions, however, have contradictory effects on the further process of economic reform and conversion. On one hand, the institutions could be the start of new market-oriented structures; while, on the other hand, these institutions often reinforce existing monopolies. Agreements to avoid competition are often signed. Banks which were founded by the military enterprises often offer grant advantages to their shareholders. These types of banks work with other commercial banks to channel or distribute subsidies from the Central Bank. The enterprises can, therefore, avoid market adaptation.

The structures of some of these corporations might be the cores of the financial-industrial corporations (FICs) which are to be artificially created from above.[7] Their intended economic mechanism should be questioned:

- What is the interest of military enterprises in transferring their expected incomes from weapon exports into civil enterprises who join the FICs? There is almost no spin-off from military to civil industry; rather, there is a spill-over from civil to military industry.

- How do these artificially created structures give incentives to develop a competitive modern civil sector?

- Will the bargaining power of these huge constructs increase without any additional responsibility being taken over by the state?

Many other questions are not answered by these elaborated concepts.

An increasing economic and political disintegration within the Russian Federation occurs while a necessary process of decentralization takes place.[8] Corporate structures compensate non-existing state policy and also include integrative super-regional elements. These elements influence the direction and dynamics of the economic transformation even if they mostly rely on arrangement processes which do not force the development of market relations but the development of barter-relations. They could, however, help to stabilise production and hinder regional separation and they could be seen as structures during a period of transition.

The phenomena of self-organisation and corporate structure development are indigenous sources of this process. These phenomena are important towards understanding the transition process as an essentially open process

---

[7] According to the concepts of industrial policy as elaborated by the Ministry of Defense (Kokoshin), by the League for Support of Defense Enterprises (Pleshakov) and other models. Cf. among others Spasut li nas FPG?, in: Krasnaja Zvezda, 31.7.1993, p.4, S. Glas'ev, Ne rynkom edinnym, in: Delovy Mir, 20-26 September 1993, p.9.

[8] Cf. A. Granberg, <u>Disintegration and Reintegration in the Russian Federation : Economic Aspect</u>, DIW, Berlin, 1993

which develops new organisational and decision making structures that eventually form their own logic and produce other development options.[9]

---

[9] Cf. H. Wiesenthal, <u>Die "Politische Ökonomie" des fortgeschrittenen Transformationsprozesses und die (potentiellen) Funktionen intermediärer Akteure</u>, Arbeitspapiere AG TRAP der Max-Planck-Gesellschaft, 93/1, pg. I-1.

Igor Musienko

Management Behaviour in Siberian Defence Enterprises

In 1992 survival was the main goal of defence industry enterprises in Novosibirsk. The best way to solve their problems (in the point of view of the time) was to return back to the previous economic conditions. The situation has radically changed in 1993. The problem of survival is now the most important problem to solve.

But, the process of privatisation, commodity market formation and change in the financial sphere are slowly creating a principally new basis of development for the defence industry. This paper will focus on some key points and key areas where changes have taken place.

1 Privatisation

At the present time, presidential decrees mainly list the number of military and nuclear industrial enterprises which are prohibited from privatisation.

During the process of preparing the presidential decrees, the directors of a number of Novosibirsk defence enterprises fought with great energy for the rights to be privatised. Moreover, the directors tried their best so that 51 per cent of the common stock or "golden share" would not indefinitely remain federal property.

One example is the nuclear industry. This industry produces nuclear fuel for Russian, Ukrainian and Bulgarian power stations and is also the main producer of lithium in Russia. Due to the great efforts of its administrators, the Government gave them permission in December 1992 to privatise the industry by allowing only 38 per cent of the common stock to remain federal property for three years.

The struggle to obtain a large portion of the shares is one of the main factors that determines the behaviour of many senior administrators in the Novosibirsk military industry. Each administrator is convinced that managers must be the owners of the controlling interest in their enterprise.

An analysis of privatisation in both the civil and military industry shows that this process has several consecutive stages.

Managers are determined during the first stage to provide a distribution of the stock which is kept between the enterprise's personnel. This step is taken to limit the opportunities for potential external investors. The arsenal of such means includes:

- establishing particular privatisation regimes by using special Government resolutions;
- introducing limits (so called "traps") into the charters of an enterprise which prevent external investors from buying large shareholdings;
- directly subsidising personnel (often at the financial expense of the enterprise) so they can buy stocks;
- limiting information about stock auctions;
- buying stocks at auctions by private firms which are owned by the enterprise managers and by other private companies which have an agreement about how to subsequently repurchase stocks.

Efforts during the second stage are directed at a step-by-step purchasing of any considerable shareholdings that are owned by managers from their own staff. Conflicts during this stage can occur between organised groups of employees who claim more rights in corporation management, the raising of wages or salaries or the distribution of the main part of any profit as a dividend. There are also many people who do not want stocks to be concentrated in the hands of administrators.

The majority of Novosibirsk military industry enterprises with permission to privatise are still at the first stage. Some enterprises learned a lesson from the experience of civil enterprises which, in 1990-1991, concluded agreements for state property leasing, set up public corporations and have now completely fulfilled the privatisation process. For example, during 1992 an enterprise, which produces hand weapon cartridges, fulfilled in record time all of the formal privatisation procedures. The combination of USSR and Russian laws and the energetic actions of its managers allowed the transfer of 88 per cent of common stock into the staff's private property.

Although some uncertainty about the distribution of shares of equity capital in an enterprise exists and although control over the movement of shares is not yet effective, managers do not make considerable efforts towards development and restructuring.

2 Market and Production Structure

Most directors of Novosibirsk military industries realise that their enterprises do not have a future in a market economy which has such production structures. However, the uncertainty of a number of important external conditions greatly prevents a restructuring process.

The level of state defence orders is an external condition that produces uncertainty. The prospect of future state orders is not clear since this order was greatly reduced in 1992. (The reduction in the Novosibirsk region was around 70 per cent.)

At the present time, the Novosibirsk defence industry mainly manufactures civil products. At the same time, a large part of capacities and personnel for military orders are maintained. Conversion programmes are therefore of a quite small scale.

An examination of the distribution of conversion credits given by the Government to nineteen Novosibirsk enterprises (excluding nuclear and aviation industries) in 1992-93 shows that 53 per cent of the credits were used to produce consumer electronics, 14 per cent to produce agricultural machines, 12 per cent to produce textile and food industry equipment and 12 per cent to produce drugs and medical equipment.

There are a few examples of large-scale conversion programmes. One atomic industry enterprise worked out, on the basis of their scientific and technological potential, a number of biomedical technologies. They began to produce medical preparations for blood clearing and different jellies. The enterprise also conducts active experiments in the field of producing equipment for the mining and food industries.

The main goal of many enterprises is the export expansion. In the past, many Novosibirsk military enterprises sold their products to their ministries for export. The prices were determined by the state price lists and many enterprises lost a considerable part of the profits. Their first experiences of directly exporting in 1992-1993 gave them optimism.

Novosibirsk defence enterprises face serious competition in the domestic market. On one hand, there is the increasing competition of import goods, especially electronics from Europe, Japan and South East Asia. On the other hand, specialised Russian civilian enterprises are strong competitors for the defence industry. Their products are much cheaper and are almost equal in quality.

The growing orientation towards export markets and the competition on the domestic market has considerably focused attention on the need to study markets before producing goods. It is in great contrast to former approaches to the problem: the choice of producing civil goods was fully determined by the technological capabilities of an enterprise; any valuations of potential markets were characteristically superficial.

Some enterprises have created marketing departments and specialised export and import companies. Brach associations have been established with the aim of studying markets and providing sales, transportation facilities and post-sale maintenance. A corporation created by electronics enterprises, SICOM, is a good example. The association, Gas Weapons, unites most Russian producers of cartridges and hand weapons.

However, the methodological and organisational basis of marketing subdivisions and their information support are at a low level. Directors of enterprises want to single out marketing as among the actual problems. They are currently looking for help to better organise their marketing programmes.

An enterprise's main duty is to fulfil state military orders; however the State, as a customer of armaments, is unable to make military orders profitable. Profitability of armament production cannot exceed 25 per cent. When the inflation rate is about twenty or more per cent each month, weapons production, in many cases, becomes unprofitable. Late payments from the state budget are also a problem. There is usually a delay in payment several months after the shipment of goods.

The problem of financing maintenance for under-loaded capacities and the problem of paying the actual wages of unemployed personnel is not yet solved. As in the past, defence enterprises can be forced to fulfil military orders by the Government. Enterprises must pay any expenses related to maintaining unloaded capacities while they wait for future orders. This is very expensive.

This is why managers consider that their financial problems are mainly determined by external factors. One external factor is when the State does not fulfil its financial liabilities. So, many managers believe that tension from financial problems can be solved by other external sources. Such sources are low-interest, preferential, Government loans, bank loans and commercial credit from suppliers.

For example, in Autumn 1993, the State refused to give several Novosibirsk enterprises preferential loans for the purchase of fuel for the Winter season. In return, the enterprises refused to pay the accounts of the state regional energy company. As a result, the energy company received a huge Bank of Russia loan. Enterprises' debts will be paid when this sum of money is reduced by inflation.

Defence enterprises pay little attention to improving their internal financial systems. Accounting and planning are organised according to previous Soviet standards and do not meet modern requirements. Accounting system implementations are used to produce tax statements, currency control statements and reports for other state bodies; they are not used to improve the financial management of the enterprise. Managers are indifferent to recommendations given to improve accounting, controlling and planning. They say, "This is theory without any practical benefits!"

Even in the present climate of privatisation and changes in the economic environment, it is difficult to determine how much time it will take to draw military enterprises nearer to the patterns of market behaviour.

Ella Amosenok and Victor Bazhanov

Specific Features of Conversion in Siberian Defence Enterprises

Official statistics on the dimensions of the military-industrial complex provide only a very perfunctory idea of the magnitude of conversion objectives in Russia and in its individual regions, Siberia in particular.

According to this data, the total workforce of the Russian military-industrial complex is presently about 6.5 million (over 30 per cent of the total industrial labour force) and the personnel of MIC research centres number 1.7 million. Russia has over 1,000 defence enterprises (one source argues that the number of military plants is as high as 5,000) and about 900 research and design facilities are oriented towards defence needs. As with the rest of industry, MIC enterprises are unevenly distributed over the territory of the Russian Federation. The majority is concentrated in European Russia (including the Urals). Siberia accounts for some eight per cent of total MIC labour and nine per cent of its enterprises. The region has several concentrated centres of defence industry - Novosibirsk, Omsk and Krasnoyarsk - and accounts for between nearly 70 per cent of the total Siberian MIC labour force. The remainder of the complex is located in other cities of Siberia. Furthermore, there are satellite towns in the vicinity of Tomsk and Krasnoyarsk which are also centred around major nuclear industry enterprises.

The core of the military-industrial complex in Siberia is made up of machine-building, instrument and electronic industries; this determines the complex specialisation in high-technology products. This argument is further supported by the fact that in Siberian cities with defence enterprises, research and development organisations of the former Defence Ministries and the Siberian Branch of the Russian Academy of Sciences operate (for the most part) by closely matching the needs of the specified industries. There are more than 50 R&D agencies in operation in Siberia. In addition to machine-building enterprises, the Siberian defence complex comprises a number of major nuclear industry plants: Novosibirsk chemical concentrate works, Kemerovo chemical plant and mineral and electrolytic chemistry factories in Krasnoyarsk.

Thus, the enterprises and organisations of the Siberian MIC are distinguished by the diversity of enterprises and industries as well as their uniqueness. With regards to the nature and extent of conversion arrangements, the region's defence enterprises can be classified as follows.

The first group comprises enterprises which are exclusively devoted to military production and have not adopted any civilian product line. They are the least susceptible to conversion and are likely to retain their military specialisation in the foreseeable future (depending on what type of military doctrine is adopted).

The second and largest group is made up of diversified enterprises with variable proportions of military and civilian output (including enterprises whose capacities are retained as mobilisation or wartime reserves).

The third group includes enterprises which continue to manufacture military products but, for some reason or other, are forced to convert to civilian production (on account of obsolescence of defence products, change in defence needs, etc.).

The identification of the above groups suggests that the conversion process is contingent on a variety of factors, not least the regional context.

The conversion process in Siberia (as in the rest of Russia) is largely dependent on the political and economic context. Any uncertainty is further aggravated by the lack of a defence doctrine and, hence, hardware and ammunition orders from government agencies. This process has given rise to a complicated and contradictory situation in the defence sector which manifests the following characteristics:

- a spontaneous increase in unfinished production of military goods and stockpiling of inventories which cannot be adapted to civilian production;
- a drastic drop in aggregate profits for defence enterprises under conversion and a decline in the rate of return, thus, resulting in a large number of loss-making and low-profit enterprises;
- the additional costs of equipment preservation and maintenance of mobilisation capacities with an ensuing rise primarily in the cost of civilian products;
- increasingly acute problems posed by uncertainty as to the size and structure of capacities to be retained for defence production. This blocks their commitment to civilian production, one controversial issue being the rational choice of product mix to replace military output ensuring the least alteration in the established technological characteristics of each enterprise;

- Siberian defence complex enterprises are more affected than those in European Russia by the breakdown of economic links, particularly with the former USSR republics. There are no narrowly specialised defence enterprises in Siberia. In many plants, the product range exceeds 50 items; the plant is a monopoly producer for most of these products.

It is becoming obvious that a further increase in the production of consumer goods in defence enterprises by following the current pattern of conversion is bound to bring about an adverse phenomena in the economy which will involve the unproductive use of material, energy and financial resources. This spontaneous, uncontrollable conversion is inherently incapable of accomplishing a major objective which regrettably was not formulated in the original concept - technological change in the civilian industries (with the exception of light and food-processing industries).

The recent reduction in defence orders (to one eighth of the 1991 level) has led to a dramatic contraction of production at defence enterprises in Russia and Siberia. Thus, estimates made by defence enterprises in the Novosibirsk region suggest that total output across the whole product range will decrease by the end of 1992 by not less than 20 per cent against 1991 while military production will decrease by 70 per cent and bring its share of total output to 11-12 per cent. The proportion of civilian products will rise accordingly to 88 per cent including consumer goods at 35 per cent. In the Omsk region, the aggregate output decline is in the range of 3-44 per cent while military goods are at 13-60 per cent.

The large majority of defence enterprises in these two regions already have less than 20 per cent of military products in their total output; civilian production is predominant in 85 per cent of enterprises. At least six enterprises in the Novosibirsk region and five in the Omsk region have the lowest percentage of defence production in their product mix (not over 10 per cent) and have already assumed civilian status for all practical purposes. A number of Siberian enterprises, however, still maintain a high proportion of defence production (over 50 per cent).

The fall in defence production is only partially offset by growing civilian output. Thus, while defence production in the Novosibirsk region dropped by 10.2 million roubles in 1992 (compared with 1990) civilian output rose by only 4.7 million roubles. This means that industry is unable to compensate for over 55 per cent of the decline in military output over the last two years. In 1991, the civilian output of the defence sector in the Novosibirsk region grew by 17 per cent over 1990 and in 1992 it is expected to bring an increase of another 18.5 per cent.

Specific Features of Conversion in Siberian Defence Enterprises

A distinctive feature of the situation is that most enterprises are incapable of making up for shrinking military contracts by boosting the production of civilian goods. This is due to several reasons.

Firstly, production assignments to enterprises undergoing conversion are still determined "from the top" as government orders fail to take into account the specifics of their production process and technological potential; this results in an inefficient use of productive capacities, personnel capabilities and involves extra costs. As a result, by early 1992, the prices of such new production items soared nearly tenfold over similar products manufactured by civilian machine-building industries; the volume of civilian production increased in money terms but not in physical units. These developments merely contribute to inflationary pressures and aggravate the financial plight of defence enterprises rather than bailing them out.

Secondly, a number of enterprises are distinguished by the absolute impossibility of adapting their technologies to any type of civilian production.

The sharp drop in military orders and the failure to countervail it with civilian output has resulted in a dramatic increase in unfinished production at enterprises in the Siberian defence complex. For example, there was an increase of 150 per cent in the Novosibirsk region when compared with 1990 figures. One fifth of the unfinished products' cost is simply written off because the products are unusable by the enterprise or elsewhere in the national economy. The situation is compounded by mounting inventories nearly half of which have already been written off.

The existing research and technological potential of defence enterprises is no longer fully utilised because of the above factors. At present, most defence complex plants are seeing the release of fixed assets which were formerly required for the manufacture of military products. By the end of 1992, these processes already left large vacant areas at defence enterprises. On one hand, the uncertainty about military contracts causes grave concern at the enterprises about the need to keep these premises and maintain the equipment. On the other hand, in the event of effective contraction of production and dismantling of equipment which cannot be adapted technologically to the manufacture of any civilian goods, enterprises have the opportunity either to use the vacated premises for an entirely different production process, to lease them to other business entities or to attract foreign capital and set up a joint venture on mutually beneficial terms.

The processes of under-utilisation of defence complex production potential are further complicated by its heterogeneous composition which combines up-to-date, high performance equipment with large quantities of obsolete stock. Thus, in many plants, over 40 per cent of (mostly multipurpose) equipment has been in operation for more than 10 years. This equipment cannot physically stand up to the workload it was originally designed to handle. As the replacement of old equipment in the context of an economic crisis and lack of funds is severely retarded, the depreciation of defence related stock continues while threatening to escalate out of control.

Individual defence sector industries have historically turned out various civilian products since their inception. Practically all consumer electronics - home refrigerators, washing machines and other electric domestic appliances - numerous types of household tools and other utensils were manufactured by defence plants. The sweeping conversion programmes of 1988-89 and the continuing uncontrollable processes in subsequent years led to a spontaneous specialisation in civilian capital and consumer goods production.

A certain change at this point can, however, be witnessed in the conceptual framework of conversion: from a mandatory switchover of practically all defence enterprises into manufacturing civilian products or consumer goods to a programme which provides for re-gearing military plants towards making new products and technologies without counterparts on the world market. These efforts draw on the results of large-scale R&D efforts and basic research and pioneering development which were originally intended to serve military needs while simultaneously developing military hardware for weapons production exports. The underlying idea of this concept is the self-financing of enterprises under conversion so that hard-currency revenues can be used to fund structural change (including civilian production) and to meet the social needs of their employees by purchasing foreign consumer goods.

This idea has already found its way into conversion programmes up to 1995 that have been drawn up by virtually all defence enterprises in Siberia.

The vast majority of Siberian defence enterprises belong to the second group which, as noted above, is characterised by a varying degree of production diversification. An important feature of defence machine-building plants is that their production capacities for military and civilian goods are technologically and territorially separate. The exception is instrument and electronics factories where the same technological equipment is used for producing military and civilian goods. Hence, there are different ways to solve problems related to changing civilian production structure and to increasing output, in particular, by using industrial premises released from military production and by using new production organisation patterns.

For example, enterprises, where substantial capacities are realised from discontinued military production and where technological change is an issue, face the problem of adapting these assets to the massproduction of civilian goods. The difficulty is that military hardware manufacture was usually associated with small lots and few customers. The transition to mass production will only be effective if high productivity is combined with lower costs for civilian products (comparable with products from civilian enterprises). This calls for an expansion of productive capacities through the addition of high-technology equipment and through a more advanced organisation of production.

The former aspect is presently severely complicated by the fact that specialised civilian machine-building industries (for example, the machine tool industry) are incapable of producing technological equipment which compares to Japanese and German standards. The production of precision equipment is especially difficult because of the long-standing technological lag in both production and R&D. (This is also further aggravated by the present crisis.) The defence machine tool industry is currently unable to massproduce technological equipment for modernising obsolete, multi-purpose equipment at Siberian defence enterprises.

The introduction of new organisations of labour involves not only a change in management and organisation of production, but the development of new functions - marketing services for effectively accessing different markets. Such steps are in the interests of both enterprises and potential foreign investors, since more information about the priority needs of Russian defence enterprises which seek access to international markets could be useful. At this junction, the MIC cannot apparently dispense with foreign investment since it could provide an initial impetus to the solution of major conversion problems.

Any foreign partner's interest in participating in the conversion processes in Siberia, for example, may be due to certain comparative advantages specific to defence sector enterprises. This includes a highly trained and skilled yet relatively low-cost labour force, availability of industrial sites - vacant, about to be vacated, incomplete construction projects (they usually have adequate utilities, transportation and access roads) or recently built/reconstructed real estate - less stringent environmental requirements compared with world standards, etc. In this sense, defence sector enterprises can offer their services on foreign markets as subcontractors for organising the production of parts and assemblies for a sufficiently wide range of final products - equipment, durable goods, etc.

For example, the basic output of aerospace, chemical and electronics industries - product materials with special properties, environmental conservation equipment, unique measuring instruments, etc. - can be of interest to foreign investors in Novosibirsk region. Co-operation with foreign firms in these fields can proceed by establishing a programme of investment funds which would serve production and R&D with a view towards enhancing the competitiveness of existing products or the development of new products.

It should also be noted that the traditional isolation of the military and industrial complex appears to have a positive aspect as well: a large number of technologies and research products which were developed in various fields of the defence complex are highly innovative and ahead of leading foreign companies. The prime objective regarding these technologies is to assess their applicability to civilian uses and to find national and international market niches that could be filled by them. Co-operation with foreign companies can make a positive contribution here, particularly in terms of hard-currency investments.

In conclusion, the final goals of the Siberian defence complex are not only associated with its own structural change in basic production and the protection of social facilities it has built up over the years, but also with the impact of these changes on the restructuring of the entire Siberian industry. If conversion results in identifying, strengthening and developing industries which formerly served defence needs but now possess sufficient high technological potential to permit the manufacture of internationally competitive products, this could resolve many of the region's development problems and promote its share and influence in the Russian economy. To develop successfully, all the specified processes need not only tax and credit benefits (which so far are insufficient), but also certain protectionist measures and strong government support in terms of legislation and priority investment policies.

Grigori Tomchin

Conversion and Privatisation: The Experience of St. Petersburg

Taking into account the specific character of Russian industrial structure and especially that of St. Petersburg, one of the main problems of the reform is the structural reorganisation of industry and, as its constituent part, the conversion of the military industry.

The problem of conversion is not just a problem of transition from military to civilian output, but the radical problem of changing the approach to manufacturing itself. All Russian enterprises (especially the military ones) were oriented to one general plan so that attention was not paid to marketing problems. The main managerial role in Russian enterprises during the totalitarian period was played by a supply agent, while the management was not concerned with the financial standing of the enterprise at all. Consequently, we now have to deal with a management culture which does not recognise market problems. Their goal is the technical grounds of the demand for output and they do not care much about their enterprise occupying a corresponding niche in the market or about the output itself. Therefore, the whole Russian industry was not concerned about finding its own niche in the world market.

Initial contacts with the world market made a sort of selection in manager's minds. The most qualified realised that privatisation was the most realistic opportunity to attract investments and to adapt the enterprise to market conditions. There is no peculiarity concerning the privatisation in the kind of manufacturing, output or equipment. Branch-specific features depict the degree of difficulty entering the market as well as the varying amount of capital outlays in the case of manufacturing reconstruction with a complete change of profile.

The abrupt reduction of military orders at the same time as state orders led to the necessity of completely reconstructing manufacturing. The fourfold reduction of military orders does not mean that you can use only one fourth of the manufacturing facilities at a given enterprise. Instead, it shows the necessity to change all the equipment and the organisation of the manufacturing process itself. The enterprise is often unable to carry out the reduced plan. It is connected with the scienctific and technological complexity of the manufacturing process and the product. The intrusion of military enterprises into the sphere of civil production forces civil enterprises to deal with structural reorganisation since they cannot compete with military enterprises in terms of quality of the staff and equipment. All of this led to a chain reaction which led to the necessity for the structural reorganisation of all Russian enterprises. This is practically impossible without privatisation under the conditions of inflation and the state investment shortage.

Privatisation schemes in Russia are undertaken on the condition of rapid privatisation so as to retain a certain social peace. This schematic is characterised by algorithms which are invented not in the silence of cabinets but in the compromises gained as the result of political struggle for reform. The first part of it is satisfying the workers psychologically or satisfying those who live according to the psychological principle: "Factories for the Workers". This principle is expressed in the system of privileges given to employees during the acquisition of stock. The State does not indulge the collective selfishness of the staff, but forces everyone to make his/her own choice by giving an opportunity to sell the stock on the free market. Besides, the staff has no right to preferential distribution of stock. At this stage, each worker confronts the choice of how to buy a bigger amount of stock. Changes in a worker's attitude towards his/her own enterprise usually leads to the purchase of only 5-25 per cent of stock intended for the staff by the management (5-30 persons). It does not depend on the scheme (variant) of privatisation chosen by the staff. Management purchases up to 40 per cent of the stock after adjustments to the market relations.

The second constituent of privatisation - so-called "cheque auctions" or "people's privatisation" - aims at several goals:

1. it makes a quotation of stock which resembles the market price;
2. it involves the financial institutions - banks, investment funds - in the process of privatisation.

In St. Petersburg, 25-29 per cent of the stock of practically each enterprise enters this type of market. This process is a distinctive peculiarity of the region, but it does correspond to Russian laws.

These two constituents of industry's privatisation lead to each enterprise practically having two relatively big owners, namely, the management (5-20 per cent) and a financial institution (also 5-20 per cent). But the specifics of the newly born financial institutions in Russia does not make them real investors in an enterprise. It can be easily explained: the investment funds have no capital, whereas the banks consider the investments in the enterprises to be of small profit in the period of inflation. Management, by having a sufficient amount of stock, does not have investment capital at all.

Thus, these two constituents do not essentially create the conditions for conversion of the enterprises (the absence of the investments). Perhaps nothing terrible would happen with one enterprise or a group of enterprises if the remaining ones were in a normal financial situation. The new owner would sell his stock to the real investor on the secondary market. But, as this situation takes place within all enterprises, we would find that unemployment would dramatically increase for the period time spent in looking for the owner and, as a result, a return to the totalitarian regime.

The idea of the third constituent of industrial privatisation was developed in St. Petersburg. This type of privatisation aims at attracting investments which are already on the primary privatisation market. In this case, the remaining 20 per cent stock packet (after the stock being shared between the staff and the cheque auctions) is sold on the so-called investment sales or investment competitions.

The criterion of the competition is not the price of the stock packet that can be scorned but the fulfilment of a concrete investment programme for an enterprise's development. In these cases, the State is initially the founder and the sole owner of the whole stock packet of the privatising enterprise. It determines the founding of the joint-stock company by the availability of the 50 per cent of the capital increases in the privatisation plan. The rights to obtain the 50 per cent of the capital is sold together with 20 per cent of the stock. It should be noted that the investor can use this right only after fulfilling part of the agreed investment programme or after fulfilling the entire programme. As a rule, these investment programmes exceed the regulatory capital by their volume by five to a hundred per cent or more. By purchasing the stock and obtaining the right to increase the capital, the investor gets up to 47 per cent in recounting to a 100 per cent capital. This percentage level is considered to be the control packet even in our wild market.

One can suppose that this method of sale will cause a negative reaction on behalf of the management since control over the enterprise is transferred to the investor. However, the property Committees confirm the programmes offered by the management since they have no possibility or time to make up the investment programmes themselves and since they purposely want social stability. The real investor, used to the long study of the enterprise and its examination, is obliged to contract the management before the privatisation occurs and during the period in which the investment programmes are prepared. This process, on one hand, takes into account the particularity of the enterprise before privatisation (the management) and, on the other hand, the possibilities for the enterprise on the market (the investor). Such a scheme makes a selection of the management. By trying to attract their own investor and by wanting to be sure of the sales victory, the management makes the programme extremely difficult and increases its volume. The only thing that can hold it back are the possible influence of the investor found. If the management fails to find an optimal investor, another investor can carry out the suggested programme and replace the management thereby guaranteeing the State as the optimal winner.

There is still a problem: how do you keep yourself away from an unscrupulous consumer or a middleman? The state makes the primary selection in the investment programmes, i.e., the participants of the sales should be the investors who give guarantees that the programme will be fulfilled. The management carries out the choice of the guarantees and inputs them in the programme.

The first experience of holding investment sales in St. Petersburg and Russia offered an opportunity to a number of industrial enterprises to enter the new market structures and to even begin manufacturing competitive output for domestic and international markets without using state credits.

At the present time, we are making an attempt to massively use the suggested scheme in St. Petersburg.

Now let us look at some questions regarding how one finds an investor and how foreign capital is attracted.

Many native and foreign companies specialise in a definite kind of industrial manufacturing, but these companies often do not have much free capital to use in developing new manufacturing facilities. Financial institutions also do not like to risk inputting into investment programmes for this or that enterprise alone. The lack of large free amounts of money means that Russian and foreign investors run a risk of investing in a small number of enterprises without getting back real profits. This risk is increased by the high speed of making a decision regarding privatisation; it does not allow the investor to make profound preparations or a system of examinations. Besides, if the stock packet is bought by only one attracted investor, the interests of management may be unbalanced and could lead to a conflict between the management and the new holder of the stock packet. Small shareholders - the staff, the investment funds or the individual buyers on the cheque auctions - could even become engaged in this type of conflict . Such kind of conflict is not dangerous to the development of the enterprise itself as the conflict can be solved within the framework of the legislative base of the joint-stock company. However, this situation could lead to a temporary lack of investments (until the matter is settled) and, as a result, could reduce the number of vacancies and other social post-effects.

If privatisation was not on a massive scale and if the market was already formed, this point would not arise; but, in dealing with a non-formed market and large scale privatisation, the balance between an investor, the management and the small share-holders could lead to social tension in the region and in Russia. The existing legislative base makes the subjects of the conflict look for various methods to solve these types of conflict which already exist at this primary stage of privatisation. These variants are not abstract schemes; they are based upon world experience. Let me show one of the most attractive ways of settling this kind of concern.

In a case where the management of a state enterprise is qualified enough to guide a given enterprise and is oriented towards market relations, an investing company may be and is usually organised. This company includes enterprise management, one or two big investors - consumers of future production and/or manufacturers of a similar product which is already competitive on the world market - and a financial investor. What is the contribution of the participants? The management gives the stock packet (purchased on favourable terms), skill, licences, patents and the right to present the investment programme, while big investors may contribute a certain know-how, equipment or marketing rights. The financial investors' share is capital.

Allocation of votes and, consequently, management rights in this company occurs without the interference of state authorities and is not specified by basic legislative documents, i.e., this type of relationship gives the members freedom in their activities. In terms of the State, they form a unified company and may take part in any security sale from the staff, from cheque auctions or from investment sales.

The formation of an investing company is preferable for the State during the early stages of privatisation because this company, in order to succeed, begins to invest long before the end of privatisation and sometimes even before the start of this process. The positive example in St. Petersburg is the privatisation of such giants of power mechanical engineering as Electrosila, Metallicheski Zavod and a meat-packing factory with 7,000-12,000 workers. One should not be worried that the structure of these sluggish monsters remains unchanged. Eventually all of these enterprises will start to divide into smaller parts during the post-privatisation period; one will not have to worry about their future.

The process taking place in Russia imposes a specific type of work on reform in St. Petersburg. The city turned out to be the only sea outlet to the civilised world (as during the time of Peter the First). The historical, cultural and architectural value of the city is also well known to the world public and to businesses. The activities of the City Reformers group and its desire to protect Russian and foreign investors by law are carried out in accordance with these objective factors. Russian and foreign experts who deal with the definite investment programmes and analysis of the situation as a whole consider St. Petersburg to be the most interesting region for investment in Russia. It is desirable to complete the above material with a good saying, "Nothing ventured, nothing gained."

Elena Denezhkina

Problems of Conversion and Privatisation in the Military-Industrial Complex of St. Petersburg

This paper reports on the initial stages of a research project on military conversion in St. Petersburg which is being carried out by the author in conjunction with Professor Julian Cooper of CREES. The project is funded by the UK Economic and Social Research Council.

Conversion of the Russian military-industrial complex entails, on one hand, a complete overhaul of management structures within and between enterprises and, on the other hand, a radical re-definition of the role of the enterprise in the wider economy. In particular, it is linked to the development of new economic mechanisms during the transitional period. One therefore has to situate conversion of the defence industry firmly within the context of the wider changes that are affecting every aspect of Russian life. Among the wider changes which have major significance for conversion are the following:

a) the Presidential edicts and directives which determine
   - the military doctrine that underlies foreign policy and
   - the approach to be adopted regarding privatisation of military-industrial enterprises;

b) the promulgation of successive laws on conversion that characterise the rights and responsibilities of enterprises and related organisations in terms of government contracts or horizontal partnerships between enterprises;

c) the rapid development of a commercial banking sector to facilitate the growth of credits for conversion programmes;

d) the partial or wholesale restructuring of scientific and manufacturing establishments in the defence sector, whereby they attempt to re-orientate themselves towards new markets and civil production while fulfilling residual state orders;

e) the socio-economic processes: the fall in the standard of living of defence sector employees may be seen as reducing the incentive to pursue conversion strategies since the rewards for technological innovation are significantly less than for general commercial activities. The development of environmentally-friendly products and processes had been high on the public agenda during the earlier perestrokia period; this has now subsided since reforms have undermined livings standards for most of the population;

f) the increasing regionalisation of the economy with decentralisation of decision-making, new horizontal transport and communications linkages between enterprises. (These transactions were previously routed through Moscow often to the extent that neighbouring enterprises would not be aware of each other's existence let alone their address or product profile.)

These factors have influenced the success of conversion. Development of horizontal links has paradoxically enabled maintenance of technological interdependencies between enterprises involved in the same manufacturing cycle, even after the disappearance of the central authorities who had previously co-ordinated them.

Given the complexity of the conversion process and the quantity and range of influences involved, it was decided to focus the research project on the North West region, specifically St. Petersburg and Leningradskaya Oblast. This paper is focused on the same region.

1 The Choice of St. Petersburg

The advantages of choosing St. Petersburg as the focus for the research are as follows.

a) the city's industry is strong in a wide range of relevant sectors, notably radio-electronics, instrument-making, machine-tools, vehicles, opto-mechanics, communications systems, medical equipment, aerospace and shipbuilding. Shipbuilding is perhaps the strongest sector of all. The city contained 70 per cent of the former Soviet Union's shipbuilding capacity; it covered all stages of the manufacturing cycle - scientific institutes, construction bureaux and assembly yards. 182,000 scientists work in the city's 440 major research centres. One-sixth of them are involved in industrial research and development and at least 30 per cent work in the military-industrial complex. The military-industrial complex also employs 40 per cent of the city's engineers.

b) St. Petersburg is a subject of the Federation, it has a coherent identity as a scientific-industrial centre and it is also a regional economy in which reforms, in terms of privatisation and restructuring, have generally proceeded more rapidly and dynamically than in most other regions.

c) The city's scientific and industrial enterprises have pursued innovative conversion programmes in a number of cases.

d) St. Petersburg's has become the main centre for financial services in the Federation. This role was recently confirmed by the Presidential edict of 19 November 1993, which restricted foreign banking operations in Russia to those of Credit Lyonnais and Banque Nationale de Paris (both banks are in St. Petersburg).

The disadvantages of taking St. Petersburg as an example of conversion in Russia as a whole include the following reasons.

a) Despite the coherence of its identity as a scientific and industrial centre, its enterprises are tied into manufacturing cycles involving enterprises in many other regions. To focus purely on St. Petersburg while ignoring these wider links, would provide an incomplete picture of how the process of conversion operates.

b) Despite St. Petersburg's pre-eminence as a reformist city, the city's accelerated privatisation programme has tended to emphasise "malaya privatisatsiya" or small service sector businesses.

A particular methodological approach (depicted in Figure 1) was developed and took all these circumstances into account. Our research is primarily grounded in interviews with managers and specialists from various industrial enterprises, scientific and project centres as well as local authority representatives persons from the local subsidiaries of the Moscow Centre of Conversion Economics, local press representatives, and staff from Goskomstat, banks, consulting firms and other relevant organisations.

During the first phase of the project (August-September 1993), 35 interviews were carried out in 18 enterprises and organisations in St Petersburg, 19 interviews were carried out for comparative purposes in nine enterprises and organisations in Ekaterinburg, 11 interviews were carried out in Moscow organisations which are concerned with the overall conversion strategy and three analogous interviews were carried out in Kiev for comparative purposes.

2 Presidential Edicts on Privatisation of the Military-Industrial Complex

It is impossible to separate conversion from the wider process of reform. Defence enterprises are often caught between contradictory pressures from above: they are largely dependent on state orders while they are also expected to comply with the overall privatisation strategy. It is with this paradox that we begin the following review of conversion via a discussion of recent Presidential edicts. This section draws on a range of interviews conducted on 24 August 1993 with Dr. Salo, Deputy Minister of the Economy of the Russian Federation and Major-General Kulichkov, Director of the Moscow Centre of Conversion Economics.

Presidential edict 1267, "On the specifics of privatisation and supplementary measures regulating the operations of defence-related sectors of industry," sent waves throughout all levels of the military-industrial complex. The edict provides answers to a number of the most fundamental questions asked by the defence sector, such as, "will the mass, uncontrolled privatisation of defence enterprises be brought to a halt?", "which enterprises of the military-industrial complex will be submitted for privatisation and which ones will not be submitted?", "will the process by which defence-related enterprises are privatised be distinct in any substantive way?" and "what is the role, rights and responsibilities of the State as a co-owner of a joint-stock defence enterprise?"

This highly significant document differs from many other edicts that give special rights to specific sectors regarding privatisation (see the edicts on agro-industrial and energy privatisation). The first point of the edict declares,

> "a cessation of the transformation of state enterprises and organisations of the defence sector into joint-stock companies and subsequent privatisation ... until such time as the Council of Ministers confirms the list of the enterprises and organisations which are not to be privatised, this cessation not to last more than three months from the moment of the edict's coming into force."

This meant that the speed and scale of the privatisation process would depend to a large extent on the contents of the list which was contained in the supplement to the edict. The list itself proved to be considerably longer than the edict itself. According to the data of Roskomoboronprom (the State Committee for Defence Industry) 28 per cent of the most important military-industrial enterprises and scientific or project organisations which are concerned with Russian defence capability were included in this list - a total of 475 organisations out of a possible 1700. The Council of Ministers was obliged to confirm the list by the end of the three-month period after consultation with the Ministry of Defence, the State Committee for Defence Industry and the State Property Agency.

This would appear to imply that, after the three-month period has elapsed (19 November 1993), the 72 per cent of defence industry organisations which were not included in the list are to be turned into joint-stock companies and, as the Russian press puts it, "launched onto the rails of the free market."

Things, however, are not that simple. As with 'not a few' of the Russian reforms, the trajectory from initial conception of the idea to its eventual realisation is so lengthy and tortuous that the original idea is no longer recognisable in the end result.

First, there are enterprises included in the list which have already been privatised in the sense that they have already been subject to a voucher-based auction. Secondly, the end of the three month period happens to coincide with the expiry date of the vouchers issued to potential buyers.

As P. Mostovago (Deputy Chairman of the State Property Agency) has commented, "Into the list which was compiled by the Ministry of Industry went enterprises which fell into a grey area between sections 2.1 and 2.2 of the State privatisation programme - between the category of 'privatisation forbidden' and that of 'privatisation subject to government approval'".[1]

The edict sets out the necessary pre-conditions for privatisation: there should be formal agreements requiring the enterprise to fulfil state orders and to observe state secrecy. These agreements are required to be drawn up in accordance with model agreements prepared by the government. Based on past experience, it is perhaps not surprising that such model agreements do not appear to be available in practice. There is, thus, no clear guidance on the respective rights and responsibilities of either the State or the enterprise concerned. The demand that the State Committee for Defence Industry should prepare such model agreements within the three-month period is of major importance in theory, but it amounts to very little in practice. Meanwhile, the State's record as paymaster has gone from bad to worse. During the first nine months of 1993, payments by the Ministry of Defence for state orders from defence industries fell short of the agreed level by no less than 330 billion roubles - outstanding debts from the previous year should be added to this figure.

In the interest of fairness, one should note that the edict does give some consideration to the responsibilities of the State regarding contract obligations. Thus, the State Property Agency, the State Committee for Defence Industry, the Ministry of Finance, the Ministry of the Economy and the Ministry of Justice are all instructed to draw up, "guidelines on State liabil-

---

1   Interview in "b", 22 September 1993.

ity regarding the obligations of state enterprises and organisations, the main operations of which are funded out of the Federal budget."

This point may be regarded as a rather hesitant step on the part of the State in the direction of supporting only those defence enterprises that are exempted from privatisation. If this is the case, it follows that those defence enterprises not included in the list (the majority) should prepare for when they will be thrown headlong into the maelstrom of the Russian market.

It still remains unclear what type of support the government is prepared to extend to privatised "P.O. Boxes" (as secret enterprises used to be termed) nor is it clear how state secrets in the form of scientific and R & D documentation held by such enterprises will be maintained. The enterprises concerned are still in the early stages of building horizontal links in the wake of the Soviet Union's collapse and, in many cases, their real or potential partners are scattered across the Federation and beyond. It is far from clear what kind of co-ordination (if any) will be available to help them adapt to the market in the face of such difficulties.

The edict seeks (at least in very general terms) to clarify these issues, but it leaves sufficiently large areas of ambiguity as to raise major doubts as to the government's real intentions.

Thus, a number of ministries are instructed to prepare, "proposals for the founding of financial institutions and guidelines for the use of dividends from federally-owned defence enterprises which have been reconstituted as joint-stock companies." The idea here seems to be that of setting up investment funds aimed at, "wholly-financed conversion initiatives, technical re-equipment, reconstruction and broadening of product ranges."

Taking into account the fact that a significant portion of city infrastructure - higher education institutions, kindergartens, hospitals, catering outlets, shops, etc. - may still be found on the balance-sheets of defence enterprises, a part of the proposed funds are planned to be set aside for, "pursuit of ecological initiatives and the maintenance of socially-important units belonging to these enterprises and organisations."

The edict's stipulation regarding directors of defence enterprises undergoing privatisation via joint-stock status represents a significant change. A director of such an enterprise must have, "a qualification certificate in accordance with procedural guidelines set down by the Council of Ministers." The Council of Ministers is given one month to prepare a position-statement on the rules for a certificate which will give the holder the right to manage a Federal joint-stock company and on the list of products items from which need to be produced as a pre-condition of such a certificate being granted. This point clearly provides the State with a control function regarding the production of strategically-necessary products in a relatively open and free market context. It is not made clear whether or not this arrangement is for the long or short-term.

A similar issue arises in a supplementary edict's stipulation. In a case where one person has acquired a controlling interest in a defence-related joint-stock enterprise, the State will then exercise special rights regarding the enterprise and its management. These rights are to remain in force for a period of not less than three years; however, a provision is made for this term to be altered by further legislation. Although the edict makes it clear that the concerned enterprise should become joint-stock and although it spells out the ownership nature, it remains unclear how much owners will actually control the enterprises and how this control will operate.

3 Secrecy

One issue any privatisation programme in this sector needs to resolve is that of the boundary between the open and closed parts of any military enterprise. The edict states that those parts of enterprises (including their documentation) which are objects of state secrecy will remain, "exclusively Federal property and will not be privatised," and can only be operated by agreement between the State Property Committee, the Russian Military Industry Committee and the Ministry of State Security. The edict further states that decisions as to what constitutes state secrecy will be decided between these ministries and they will reach an agreement as to which parts of an enterprise may be privatised.

If such strict measures are necessary to preserve secret aspects of otherwise open enterprises, it is not clear how this ties in with the list of secret enterprises mentioned earlier. According to that list, many enterprises with secret components ranging from 10 per cent to 80 per cent will be left to their fate on the open market. According to P. Mostovago (interview cited earlier), this category may include 100 per cent of the enterprises in the military sector and it raises questions about how the policy will work in practice. An attempt to find a compromise between the interests of the State and those of entrepreneurs lies behind the policy. As Mostovago argues, the latter will be able to use documentation in the enterprises which had been developed under State funding so long as they make an acceptable agreement with the government on how the documentation will be used.

This implies that a failure to make or keep such an agreement will mean an annulment of the right to use the documentation concerned. This does not, however, appear to extend to innovations which have already been developed and may be entered into production. It is difficult to know how the flow of information will be controlled and what organisational or economic mechanism would be sufficiently robust to prevent a post facto leakage of secrets by a joint-stock company. If such a mechanism were to be developed, it would very likely be expensive and involve the establishment of a new form of inspectorate.

The Law on State Secrecy passed not long ago by the Russian Supreme Soviet is, therefore, of vital importance; however, this fact appears to have missed the attention of commentators in both the East and West. It is almost impossible to effect the transformation of ownership status of military-related enterprises or open them to full participation of Western investors without an effective mechanism to defend secrecy of scientific innovations in the military field. (Both events are happening even as the legal position becomes more ambiguous.)

The law is ineffective according to the opinion of specialists from the Directorate of State Security because it is unworkable without a whole range of supporting legislation which does not yet exist. The enterprises still need the line to be drawn between open and secret production, but the traditional (post) soviet question remains: who are the judges?

The edict states that shares in military enterprises may only be bought at specialised inter-regional stock exchanges. The effectiveness of such a system hangs in no small part on the degree to which the military-industrial complex can be re-integrated for market conditions. The scientific potential of the sector is ultimately dependent on this factor. The success of such trading will depend on the availability of accurate information on the value and performance of the joint-stock companies concerned (of which there is little sign at present). Only then can useful cross-regional (and ultimately cross-republic) horizontal alliances be formed between military enterprises so as to finally enable them to compensate for the loss of co-ordinated strategy over the last two years.

If the military-industrial sector succeeds in regenerating itself with the rest of the (potentially) productive sector of the economy through such market mechanisms, it is likely that a number of larger groupings will be formed in the process in which the controlling shareholdings will be concentrated in relatively few hands. The more far-sighted investors (both foreign and local) are already emerging to take larger and larger stakes in the concerned enterprises.

However, these tendencies will not have the desired effects unless the process is handled in a more systemic fashion and is underpinned by appropriate and comprehensive legislation with appropriate control, regulation and co-ordination mechanisms, on one hand, and, on the other hand, a sufficient flow of information to support real rather than counterfeit market mechanisms. Russia needs to emerge from this complex transition with an economy that is built, not left to happen in a haphazard or contradictory way.

Economic instability, inflation, rising energy prices and the resulting extraordinary inertness of much of the military-industrial complex is putting a brake on effective conversion projects and the establishment of efficient joint-stock companies. Inflation puts up materials and energy prices after contracts have been entered into by enterprises and it leads to extra costs being added in pre-emptively which, in turn, contributes to inflation as well as to the already vast sums of inter-enterprise debt. In the meantime, the value of state orders, upon which the military enterprises are still significantly dependent, continues to lag well behind inflation.

In such circumstances, one may well ask whether the enterprises concerned can afford not to abandon military production as far as possible. There may be incentives to use their often uniquely designed equipment to manufacture simple, crude household goods for which the market demand remains high (although increasingly vulnerable to imports), but even this assumes that production lines can be adapted for civilian use. In some cases this is clearly not practical.

4 Presidential Edict, "On the stabilisation of the economic position of enterprises and organisations of the defence complex, and measures to guarantee fulfilment of defence orders."

This edict puts forward specific measures regarding government responsibility for funding conversion programmes. One of the measures is to set up an inter-departmental committee which is charged with analysing existing conversion programmes, the list of priorities, how these fit in with regional needs and how these fit in with military requirements until 1 December 1993. On the basis of their analysis, the Council of Ministers and the Central Bank are instructed to offer credits for specific projects and to provide regular support to successive stages of project business plans.

Several factors will influence the way in which the credits foreseen by the edict will operate. The edict recognises the degree to which a fixed, guaranteed, profit margin on state orders would help economically stabilise the enterprises during the current period of inflationary transition. Such a measure would stabilise relations between the State and enterprises and between enterprises at different stages of the manufacturing cycle who might otherwise pass the costs between themselves.

The level of this fixed profit is not yet known or decided, although the council of ministers were directed to set it at a level which would allow for future modernisation of the equipment involved. In other words, the level would be linked to planned financing of the life-cycle of technical innovations in a military context. We are again faced with the questions of whether those charged with providing this carefully planned approach will have the competency necessary to carry the job out successfully and what the implementation and control mechanism for the edict is likely to be. This is a continual weakness of the system, even where an edict is well thought-out and appears to meet the demands of the situation.

As a further incentive to fulfil state orders, the edict states that from 1 January 1994, military enterprises are allowed to reimburse costs of defence production by using labour costs as the unit of compensation. It stipulates that this reimbursement must not exceed eight times the average minimum monthly wage (nine times in the case of nuclear weapons manufacturing). Given the scale of the real costs involved in terms of energy, materials, etc., this can hardly be regarded as an incentive.

The armaments procurement programme up to the year 2000 (to be developed by the Ministry of Defence by the end of 1993) is declared to be the basis for funding non-privatised defence industry enterprises; this programme will therefore be the main basis for strategic development in both business and technological terms.

5 The Context of Privatisation and Conversion

Significant cut-backs in terms of military orders, continuing uncertainty over the shape of the new military doctrine (as a guide to future orders) and the accompanying strategy of sudden mass conversion has led to an average 52 per cent fall in St. Petersburg's industrial output with military output falling by an even larger percentage. The suddenness with which finance was withdrawn had a particularly serious effect on those sectors characterised by a long manufacturing cycle, such as, shipbuilding. (The nuclear cruiser "Peter the Great" was left standing idle for months on end in the Baltiysky shipyard despite being 90 per cent completed.) As the Director of the Moscow Institute of Conversion, General Kulichkov, commented, "in this jubilee year of the Russian fleet, not one single warship will have been launched"[2].

The collapse of military orders has undermined whole sectors of technological innovation, notably rocket and radio-technical systems, spy satellites, self-navigating warheads and submarines. The result has been to push a whole range of enterprises into drastically altering their product range, area of specialisation and, consequently, their organisational structures and technology.

The privatisation of large enterprises in St. Petersburg began in November, 1992. Among the best known examples were Kirov Zavod, Baltiysky Zavod, Zvezda, Izhorsky Zavod and the Turbine Blade Zavod. The processes of privatisation and transformation into joint-stock companies took place during a period of economic, social and political instability in both the city and the country as a whole.

Although industrial employment fell more slowly alongside industrial output, it fell at an average rate of 3.5 per cent per year. The decline was much faster in science and technology with large numbers of people leaving to join newly established co-operatives or other forms of small businesses (usually commerce or service businesses). The appearance of the new private sector helped usher in the period of "small privatisation". After a slow start in early 1991, privatisation in St. Petersburg had accelerated to the point where between January and August 1993, the city's property fund had sold or was in the process of selling 1,382 enterprises of which 1,032 were municipal property (mostly local shops and services) and 350 were federal property (mostly large enterprises). Privatisation of housing occurred even faster with 103,000 flats being sold in the first eight months of 1993 (16 times more than in 1992).

Privatisation and the growth of new sectors of employment in St. Petersburg (a city which enjoys a high level of foreign investment) did not begin to compensate for the potential employment implications should the city's military enterprises collectively disappear. Industry (specifically military industries) remained the main employer in the city.

The military enterprises were also paradoxically essential to the population in terms of output of consumer goods. These enterprises were able to manufacture consumer goods on a larger scale than smaller co-operatives. Unfortunately, the opportunity provided by the slow development of the private

---

2   Interview, August 1993.

manufacturing sector was not made use of. Large enterprises competed with each other to obtain conversion related credits and, in the process, generally opted for whatever product suited their existing production lines rather than what would best satisfy public demand. This occurred at a time (late 1992 - early 1993) when, despite inflation, the real wage of the public expressed in US dollars began to rise and brought with it a considerable influx of imported goods at increasingly acceptable prices. This filled a gap which neither the new private sector (oriented towards services rather than manufacturing) nor the large state enterprises (unused to thinking in terms of market opportunities or flexible mass production) had been geared up to fill. As a result, manufacturing of non-food products fell by 27 per cent in 1992-3. Conversion had definitively failed to deliver what had been claimed for it from the mid-1980s onwards.

In fairness, it can also be argued that falling demand for the kind of products the defence enterprises were best equipped to produce was a function of the squeeze on household budgets occasioned by price liberalisation in 1992. Although the average family spending the majority of their earnings on food might buy some small imported goods, they were unlikely to buy refrigerators, televisions or vacuum cleaners - product areas in which the defence enterprises were strong.

Many enterprises became loss-making once deprived of government financial support. In 1993, 11.5 per cent of enterprises in St. Petersburg and 15 per cent of enterprises in Leningradskaya Oblast were making losses, an increase of two per cent on the previous year in each case. The actual scale of the losses increased 12.5 times. Output in the city fell by 21 per cent between January-August 1993; it fell 16 per cent the previous year. Output of military enterprises as a proportion of this total fell by a further 2.4 per cent and demonstrated the failure to compensate for falling military orders by switching to civilian production. Indeed, output of heavy goods for civilian markets - tractors, agricultural machinery, excavating equipment, plastics - fell by almost 50 per cent during 1993.

Ecology was perhaps the only area in which the city's industry performed at least relatively well and then only as a result of its more general failure. Due to falling production and the effective bankruptcy of a whole range of enterprises (certain of which had been closed for environmental reasons) in the city and the surrounding Oblast, the figures for atmospheric emissions for the first half of 1993 were down 14 per cent compared with the first half of 1992. However, this lower figure still represented 80,000 tonnes of emissions. There is no space here to consider the long-term problems of toxic fluid emissions and contaminated land.

6 The Financial Crisis in the Defence Sector

State enterprises in St. Petersburg which already suffer from lack of orders have seen their finances deteriorate rapidly as a result of a growing crisis of inter-enterprise debt. On 1 July 1993, the total debt for work already carried out by the city's industrial and construction enterprises stood at 143 billion roubles; late payments (unscheduled debts) represented 29 per cent.

This has led to wholesale impoverishment of the scientific potential on which the city and its representatives continue to pride themselves. Centres such as the Yoffe Physical-Technical Institute, the Institute of Materials, the Krylov Institute and the Vavilov Scientific Centre can be regarded as a scientific resource of the world and as a national resource. Among those working in these centres are Lenin and Nobel prize winners as well as the brightest and best of the Academy of Sciences. Here we should note that scientific centres are divided between those of the Academy, higher education, sectoral institutes and factory-based research centres. Spending in the industrial scientific research complex (known as NIR) is broken down into basic research at six per cent, applied at 34 per cent and prototype (razrabotki) at 56 per cent.

More than 243,000 people (12 per cent of the population) of St. Petersburg worked in science and science-related services. They were divided between 440 state scientific organisations, including 212 scientific research centres (NII), 43 higher education institutes, project and fundamental research centres and 23 scientific research and prototype divisions of larger enterprises.

Privatisation in the scientific sector has tended to be confined to smaller organisations. The basic pattern of employment within the sector has not been significantly changed by the appearance of new organisational forms (which now employ 44,300 persons) including small joint ventures, scientific technical co-operatives, consulting and information firms. However, there has been a strong tendency for specialists to take up parallel jobs in these newer, smaller centres while continuing to remain an employee of a larger state enterprise. According to 1993 survey data collected by the University of Economics and Finance, as many as 60 per cent of scientific specialists and teaching staff were earning extra money in this fashion.

According to Professor Gordeev, Deputy Director of the Yoffe Institute, there is a desperate need to clarify science policy, the financing of scientific institutions in the Federation and the re-establishment of the role and importance of science in the economy and society:

"At the very least, there needs to be proper management of the existing funding arrangements, so that funds which have been agreed are actually paid, and on time. Those scientific centres which by their nature will always be dependent on government funding have no other sources of income to fall back upon, apart from revenues from joint projects with Western partners, which cannot be expected to provide the basic funding of the institute. . . The government must decide what it wants. Scientific institutes like ours have long been part of the European and world community into which the rest of the country is now trying become integrated. If we cannot survive in the present order of things in Russia, then who will be Russia's representatives in the world at large - will it be just civil servants and import/export traders? Or will it be our scientists, designers and engineers, who are already working with their Western counterparts on equal terms?"

A similar position is found across a range of similar institutes, each one a leader in its particular field. According to the First Deputy Head of the Federal Security Directorate for St. Petersburg and Leningradskaya Oblast, if an institute such as GSPI Soyuzproektverf were to change its profile, the country would lose the only organisation capable of designing shipbuilding installations. This is an area of strategic importance for the Russian Federation since investment in this area has been concentrated for many years on shipyards which are now on Ukrainian territory. Equally, if the scientific institute ZNII Morfizpribor were privatised, the future of the country's entire research capability on military hydroacoustics would be jeopardised.

The job of directing such institutes has changed dramatically since the collapse of the Union. The institutes are now obliged to seek out new horizontal links and new partnerships to replace the old co-ordinating bodies; they must often do this without adequate information. At the same time, they risk losing their intellectual capital - their only long-term asset - and their attempts to do this are undermined by increasing financial uncertainties, even regarding funding which has already been ear-marked for them by government.

7 Problems of the Privatisation Mechanism: the Case of Kirov Works

The degree of risk run by management in taking radical decisions is central to the privatisation process. Privatisation is proceeding in the absence of any legal mechanism for securing the rights and duties of shareholders. Privatisation programmes are determined in traditional top-down fashion by senior management, while the workforce typically looks on with a scepticism tempered by fear of subsequent rationalisation and restructuring. Worker scepticism and passivity renders the "works collective" a controlling portion in the share of capital largely irrelevant in many cases. According to reports in Nevskoe Vremya (14 September 1993), senior management in the conglomerate Zvezda were not only waging an open struggle amongst themselves for a larger portion of shares, but they were also covertly using particular workers as a means of acquiring the shares. (Under this scheme, several workers had each applied for five million roubles worth of shares via five hundred vouchers.)

In the conglomerate Polymer, one ordinary worker cashed in vouchers for shares to the value of four million roubles. Representatives of the investment company Tziti (Petersburg), after having been refused permission to acquire shares in the Kozitsky conglomerate, declared they would nonetheless be able to buy the shares (second-hand) with the end result being complete estrangement between the management of the conglomerate and its new owners.

The case of the gigantic Kirov (formerly Putilov) works is inevitably central to any account of either privatisation or conversion in St. Petersburg. It is worth discussing this oft-cited example in some detail.

Until 1988, Kirov Zavod, one of the largest enterprises in the Soviet Union, was best known as a manufacturer of tanks, armoured vehicles and artillery pieces. Five years later it ceased almost all military-related production. Its better-known civilian products are the Kirovets tractor, other agricultural machinery and vehicles, turbine systems, purification equipment, road construction equipment and various types of both large and small batch engineering. Consumer products include the mini-tractor K 20 and exercise equipment.

Disputes over the allocation of shares in the event of privatisation began in April 1991. Kirov Zavod was one of the Union-owned enterprises said to be going down the route of "spontaneous" or "nomenclature" privatisation - a procedure whereby the directors of large state enterprises take advantage of ambiguities in USSR legislation to free themselves from ministerial control and operate as if they were the sole owners of an enterprise. The director published in the enterprise newspaper a set of proposals regarding the allocation of shares. These proposals were interpreted by worker representatives as excessively biased in the favour of the director himself. The worker representatives eventually formally sought the support of the then Leningrad City Executive Committee to confound the director's plans. In the event, these plans were overtaken by the collapse of the USSR and in November 1992, Kirov Zavod was registered as a joint-stock company (AO) - the first stage towards privatisation. The Kirov Zavod AO had a basic capital of 1,086 million roubles and 25,000 employees.

These figures did not represent the whole of the original enterprise. A series of subsidiary firms were also established in which the parent firm Kirov Zavod AO had a controlling interest in almost all of them. The most important of these subsidiaries were the Kirov Zavod concern (of which the AO held 60 per cent of the basic capital), AO Intercarbon (50 per cent), AO Paritet (38.5 per cent), TOO Nevakar (54 per cent), TO Keran (91 per cent), TOO Prin (50 per cent) and others. During the period of 1991-2, Kirov Zavod also acquired interests in (among others) the following firms: Promstroibank, BAO Traktorexport, AO Stroidormashservis, AO Baltcontainer and the St. Petersburg Fund Exchange.

As far as the financial state of Kirov Zavod was concerned, the available information was not particularly encouraging. The privatisation value was based on the existing balance of the first half of the previous year. Expert opinion held that it would be a good many years before investors saw any dividends: the relationship between actual net profit (after tax and other necessary payments) and basic capital was only 32 per cent. A major technological overhaul of the plant was needed and it stood to pay large sums to the State in compensation for ecological damage, in particular for oil-product effluents.

The Kirovets tractor, one of the mainstays of the plant's converted product range, has sold poorly; the costs of energy, materials and parts involved in its manufacture are rising all the time. In addition, its design was flawed to the extent that its excessive weight damaged the composition of the land on which it was used. It was decided to cease production of the tractor altogether.

These factors taken together have led to a financial crisis at Kirov Zavod. Despite the financial crisis, sales of Kirov shares on the inter-regional share auction have not done so badly while demand has outstripped supply. Traditional (soviet) ideas about what constitutes value would thus appear to retain their hold on the perceptions of potential investors (perhaps alongside a preoccupation with the real estate rather than the productive assets of the firm). For whatever reason, Kirov Zavod may be said to have received attention quite disproportionate to its value as compared with other enterprises in the city. What this demonstrates very effectively is the weakness of the current approach to privatisation in terms of establishing a mechanism to increase efficiency and new product innovation.

8 Conclusions

1. The enterprises and scientific institutes of the Russian military-industrial complex are in an economically extremely poor position; the more privileged establishments even face a very uncertain future.

2. The absence of any clear organisational or economic mechanism to assist the switch to civil production in a period of rapidly declining defence procurement and the consequent failure of most enterprises to adjust to the new conditions reflects not only the weakness of the legislation on conversion from 1989 onwards, but also a lack of political will and administrative coherence at government level.

3. Economic and political instability, coupled with a lack of managerial flexibility on the part of the defence enterprises, has led to a situation where the sector is patently not capable of survival in the private sector. Privatisation is being pursued with little consideration of the likely outcomes. Widely cited examples of conversion and privatisation, such as that of Kirov Zavod, demonstrate little in the way of positive achievement and much in the way of what can go wrong in current Russian economic conditions.

4. The visible collapse of Russia's scientific institutions is of more long-term concern. These institutions are unable to survive on the resources promised to them by the government, especially when the payment of these funds is often sufficiently delayed as to be rendered valueless through inflation. Under-funding, demoralisation, fashion and the infinitely greater financial rewards of the commercial sector have bled many of Russia's finest institutes of their younger specialists. If such institutes are allowed to decay beyond a certain point, they cannot be re-constituted and their loss will have grave implications for Russia's long-term future in the advanced industrialised world.

5. Recent Presidential edicts on the privatisation of enterprises undergoing conversion attempt to clarify the position, but contain some potentially damaging ambiguities. For example, the status of the list of enterprises not to be privatised may lead to further confusion and instability in the sector. As with earlier attempts to regulate conversion, there is a lack of coherence about how process is to be co-ordinated, influenced or monitored.

6. At the same time, there is a growing tendency among both enterprises and scientific institutes to pursue their own strategy of adjustment. Some

examples of successful adjustment are beginning to emerge and suggest that at least part of the sector will survive the crisis of transition.

Tarja Cronberg

Enterprise Strategies to Cope with Reduced Defence Spending - the Experience of the Perm Region

This paper deals with the experiences in defence conversion in the Perm region during a period from November 1991 to May 1993. Perm ranks in terms of the absolute number of employed in military production and in terms of the share of the defence complex of total employment as one of the ten top regions in the former Soviet Union [1]

The experiences of Perm region have been studied at the enterprise level by interviewing defence enterprise managers and chief engineers. Interviews were carried out in November 1991, March 1992, November 1992 and February 1993. A third of the thirty military enterprises in the region have been interviewed.

The theoretical framework of this paper is based on an analysis of the competitive advantages between military industries in the conversion process. The analysis of Michael Porter, adapted to military conversion by Hilton, forms the starting point. Strategies applied by the enterprises to bridge the gap between military and civil production are identified. Conclusions are drawn in terms of the supply and demand factors dominating the conversion scene in the Perm region. These are related to possible policy instruments in order to enhance the potential for conversion in the region.

1 The Question of Assets

"The absence of obvious civil applications does not mean that they do not exist, but to be developed they require an organisation to play the role of translator to foresee possible applications for advanced technology in unlikely areas. Even to identify possible uses does not mean that a market exists or that it will yield an adequate level of profitability within the time horizon of private firms". [2]

---

[1] Cooper, J.. The Soviet Defence Industry: Conversion and Reform, London, 1991, p. 19

[2] Udis, B.. From Guns to Butter: Technology Organisations and Reduced Military Spending in Western Europe. Ballinger Publishing Co.: Cambridge, Massachusetts. 1978, p. 334

Hilton points out that it is not appropriate to see the defence industry as a uniform entity. It is not an entity consisting of identical enterprises operating at the same optimum output using identical resources. These enterprises are of varying size and they use many different technologies in production and in their final products. They operate in different organisational structures ranging from the single-product small firm to the large, multi-product conglomerate.[3] Hilton's own proposal is to look at industrial groupings, also within defence, defined by the assets they deploy rather than the products they sell. These groupings can then be used as tools to analyse the conversion potential of an enterprise and as a means to aid in the selection of strategies. His categories consist of competitive assets, efficiency assets, complementary assets and specialist assets [4]

- Competitive assets:
  these provide enterprises with a competitive advantage on the demand side, a differentiated product.

- Efficiency assets:
  these provide enterprises with a competitive advantage on the supply side to enable them to drive down costs.

- Complementary assets:
  these are not core to the business but enable enterprises to deploy other assets effectively, i.e. they give access to economies of scope.

- Specialist assets:
  these are core to the business and enable the enterprise to exploit the division of labour to its fullest possible extent.

In view of these assets, Hilton refers to Gilmore and Coddington's analysis [5] of a number of conversion case studies and suggests the following relative deficiencies in defence firms desiring to convert or diversify:

1. Commercial marketing skills are absent (a lack of an efficiency asset)

---

3   Hilton, pg. 11

4   Hilton, B.. Defence Conversion or Diversification: East and West: An Overview of the Literature and the Arguments. Management Research Papers. Templeton College, Oxford. 1993, p.11

5   Gilmore, J.S. and Coddington, D.C.. Defence Industry Diversification: An Analysis with Twelve Case Studies. US. Arms Control and Disarmament Agency, 1966

2. Cost base is too high given gold plating (high use of complementary assets)
3. Technology is too sophisticated (excessive emphasis on competitive assets)
4. Engineering and production is too close together (the deployment of competitive assets given precedence over efficiency assets)
5. People find adjustment to the different challenges of the commercial market
6. Unappealing (specialist assets are competitively, not efficiency, oriented)
7. Good commercial management practice alien, e.g. tight cost allocation and control (low investment in efficiency assets and high investment in complementary assets)

Hilton points out that in command economies, market size and relative development cost were not an issue if the state had agreed that production should occur. On the other hand, the inadequacies of the Soviet distribution system and the resulting shortages forced enterprises to build up a store of complementary assets to compensate for the distribution of unreliability. This overloaded the cost base and assured the effectiveness of the military-industrial complex, but it did not necessarily assure the enterprises' or the economy's efficiency.

Defence enterprises in the former Soviet Union produced consumer goods of higher quality and more reliable supply than those available from outside the military-industrial complex. This has been used as evidence of the conversion or diversification potential of the Soviet defence enterprises. Hilton points out, "producing products such as washing machines, televisions or the apocryphal titanium alloy wheelbarrow at marginal cost from inefficiently deployed complementary assets hardly qualifies as realistic diversification" [6]. This is particularly true if these products are to compete on the future European market.

Hilton is also generally critical of the general hopes attached to conversion, particularly related to high-technology R&D as a competitive advantage on the market place. He points out:

"What should be evident from the foregoing is that while it may produce a competitive asset this can only be turned to competitive advantage with the

---

6   Hilton, p. 20

right efficiency assets made effective by competitively priced effectively deployed complementary assets whether owned by the enterprise or not. One should not presume that competitive advantage, and therefore possible diversification opportunities, are to be found solely through ownership of technologically competitive assets. The process that creates such concentrations of competitive assets also has a tendency, unless positive corrective action is taken, to accrete excesses of complementary assets and excrete sufficiencies of efficiency assets" 7

The strategies of the Perm enterprises in the following paper are studied with these views in mind.

2 The Hopes of Dual-use Applications

The most immediate direction for a military enterprise looking for conversion opportunities is in dual-use applications. Some products and processes are, at least with minor modifications, applicable both within the military and the civilian sectors. As the secrecy requirements are partly waved (as was the case in the Perm area in 1992) it would be obvious to expect that a number of military products, know-how and even equipment would find civil applications. Given the gap between the technological level of the military and civil production as well as the distortion of prices (military products are often priced much lower than the actual production costs) this approach would seem both feasible and immediately accessible.

The Perm area with its concentration of aircraft engine design bureaux and production enterprises particularly attached great hopes to the civil use of these engines. An immediate dual-use approach, however, turned out to be difficult due to the differences between civil and military requirements for aircraft engines as well as a decrease in the world market for civil aircraft engines. In addition, aircraft producers in Russia are not looking for Russian engines, but for western technology (Pratt and Whitney). Engines from the former Soviet Union have had difficulties in meeting international environmental and safety requirements. A more long-term project has currently been designed in order to develop competitive civilian engines in the Perm region.

Apart from this obvious example, a number of absurd illustrations have circulated regarding dual-use during the early phases of Russian conversion. (The use of composite materials is one example with titanium spades being

---

7   Hilton, p. 20

an often cited example.) We encountered these kinds of dual-use examples in our study in Perm. Experiments, which used powder metallurgy in the production of toys, were made in a material research institute in Perm. When demonstrating a rabbit made of titanium powder, the director of the institute pointed out that the advantage of this new product would be that it would never wear off. When asked about the price of the product the director consented that it would be expensive perhaps even a thousand dollars.

The dual-use search strategy been successful in only one of the companies we interviewed in Perm. The company experienced a reduction of military orders from 10 million to 200 thousand rubbles between 1991 and 1992. Before conversion, the company had a fifty-fifty share in military and civil products and was organised within the military industrial complex. Today, the company, working 100 per cent on the civil market, has been able to retain its work force, is paying higher salaries and has even been able to raise salaries more proportionally than other companies within the military-industrial complex.

The firm's know-how is based on the production of active coal from coal or home-made charcoal. The dual-use strategy has consisted of exploiting this know-how to make household filters for water purification and respiratory protective devices for industries with dust problems. Conversion has been accomplished without outside financing or assistance. The company has initiated the search for new civil networks themselves and has not used the branch associations which have emerging from the foundations of the former military ministries. The managers of the company are young and well-educated. One manager of the enterprise characterised the search process as, "learning-by-producing" and the quality as, "by the one the enterprise knows how to produce". While military production was characterised by an order from the ministry and co-operation by association with an "Active Coal Institute" to develop new products, today, the company does everything in-house.

3 The Technology Gap

Another strategy for seeking new business opportunities may be called "closing the gap". The gap is the difference between the technological levels between military and civilian production. Today, the differences in Russia are no coincidence. The first and foremost explanation is the priority traditionally given to the military sector, which enabled shortcuts in the administrative and bureaucratic system. Not only was the Politburo of the Communist party involved, but the leaders from Stalin onwards have even personally participated in the emergence of military technology. The defence industry (as opposed to civil industry) has had access to raw materials and other supplies. Design bureaux working within the military have had access to prototype production facilities and have been able to make experimental models and tests before production.

Quality control also differs between the civil and military sectors. In the latter, designs and prototypes are carefully finished and tested. The technical documentation is checked and the personal responsibility of the chief engineer in the design bureau is enforced. Barriers to the transfer of know-how have been bridged by giving institutions the power to co-ordinate efforts or by setting up special agencies for weapon development programmes (ballistic missiles or nuclear weapons). Thus, the larger R&D institutes have, in fact, functioned as something like prime contractors in the United States.

Finally, the achievement of the "world level" has created competitive pressures within military production. "Equivalent to the world level" has been the quality mark for technology in the Soviet Union as it was during the arms race [8]. Since it was difficult for the planned economy to decide on a desirable level of technology, "the world level", "the world standard" or the "world's highest standard" constituted (because of a lack of anything else) a quality indicator for technology. This also applies to conversion (see, for example, the Russian government's programme for conversion).

The lack of supplies, priorities, competition and necessary material conditions have, in general, made it difficult for enterprises in the civil sector to reach the world level and to introduce new technology. Or, to quote Zaichenko:

"The realisation of scientific and technical achievements in our country is complicated primarily by the fact that we do not have the necessary material conditions for the introduction of new technical equipment. The majority of enterprises do not have the necessary experimental and testing base and through their own forces cannot provide for high level of organisation

---

[8] Popper, S.W.. The Prospects for Modernising Soviet Industry. Rand Corporation: Santa Monica, 1990 and Cronberg, T.. The Price of Peace: Military Conversion on the Enterprise Level in Russia. Technology Assessment Texts Number 10. Technical University of Denmark, Copenhagen, 1992

or work for introducing new technical equipment within short periods of time". 9

It should be noted that the above applies not only to civilian industries in general but also to civilian production in the military-industrial complex. Even if the same engineers work in the same factories, the technological level and the quality control procedures within the enterprise itself will differ.

In this situation, it is not astonishing that many of the scientists and engineers who are looking for new directions focus on this gap. Using military technology - for example, advanced composite materials or military testing equipment - should give the civil sector an advantage. Scientists and engineers are willing to lower the quality of military technology while, at the same time, expecting to increase the level of corresponding civilian technology.

We discovered during our interviews in Perm in March 1992 that this approach was the main direction of a search conducted by a number of the military companies. Design bureaux, which had previously worked on missile engines, artillery or steering systems, now contemplated producing glass-fibre pipes for the chemical or food industry. Composite materials research institutes, which had previously worked in space technology, looked for consumer composite products such as, canoe paddles, crutches and other similar products which require more than normal strength. The chief engineer or the scientists in charge had usually worked out a specification which described the technical details of a possible product. In our March 1992 interviews, such product descriptions were devoid of any market oriented information. The expectation was an automatic demand since better quality would be available. In November 1992, there were a number of alternative efforts to try to access the markets in, for example, glass-fibre pipes for the chemical industry.

This "closing the gap" approach is best described by one of the managers of a new, small company which has emerged from the military industry and has contemplated producing machinery for the production of artificial surface materials. According to the Director, they were producing a high quality product even though the civilian representatives needed a lesser quality.

---

9   Zaichenko, A.. Risk and Independence of Innovative Activity. In Voprosy Ekonomiki, Number 1; January 1988, p. 41-51. Also in Soviet Union Economic Affairs. 12 May 1988, p. 27-33

("Customers said, 'thank you, we only need this level'"). When confronted with the fact that maybe the customers do not need this better quality the director answered, "We know they need it; there just not willing or able to pay".

4 Planned Conversion

The early Gorbachev conversion effort in 1989-1990 has been generally deemed a failure. A number of enterprises were told to convert, they were told what to produce and given, without any further notice, the appropriate resources to initiate new production. The initiative has been criticised for its lack of serious preparatory effort and research work prior to the political decisions, for its lack of financial resources and for the absence of legal basis for conversion.[10] Others have reported that only a fraction (50 of the 500) defence firms which were to be converted were actually ordered for full conversion while only a few (five or six) were actually converted. [11]

The reality is more sophisticated. A number of the Perm enterprises engaged in conversion efforts in 1992 had heard of conversion for the first time within the context of the Gorbachev initiative. One of the enterprises interviewed, a composite materials research institute, was actually among the 500 to receive the conversion order. The company started production of medical syringes in facilities previously used for aerospace research. The product consists of the needle itself; the total system for injection is produced by others.

Their expectation is to produce 600 million needles in the course of a couple of years. The plan of the company, which has become an independent small business (it, however, still belongs to the materials institute) is to start production of whole injection systems, as their market assessment (November 1992) shows that the producers of the whole system have a larger profit margin. It also seems that the company has had some user contact. In November 1992, the technical characteristics of the product were no longer solely defined by the original order to convert, but by the clients as well. To-

---

10 Izyumov, A.. The National Experience of the USSR. Proceeding at the United Nations Conference on Conversion: "Economic Adjustments in an Era of Arms Reduction", Moscow, 13-17 August 1990

11 Ballentine, K.. Soviet Defence Industry Reform: The Problems of Conversion in an Unconverted Economy. Canadian Institute for International Peace and Security. Background Paper, July 1991

day, the company is looking for international contacts, particularly in France and Italy. A preferred arrangement could be a joint venture where the company would get access to equipment in exchange for products.

Despite the fact that the company has access to know-how and a market for Russian products currently exists, the company is facing technical difficulties in producing a product technically competitive with corresponding European quality.

5 In Search for Demand

Military-industrial enterprises in the former Soviet Union never had to define market needs or to study demand. Consequently, the enterprises do not have any traditions or knowledge about market assessment. Most enterprises had a planning department which worked together with the central planning in Moscow. In March 1992, a number of the interviewed enterprises had recently (last week, last month) established a marketing department, often by renaming the planning department. Between the interviews conducted in March 1992 and the ones in November 1992, there was also an increasing awareness of how important marketing and market assessment are in conversion proposals.

The lack of marketing capabilities does not necessarily mean that the companies are only searching along the technology push lines as identified above. Some demands are generally acknowledged and visible to everyone. In fact, everything seemed in 1992 to be in demand, particularly food and housing. A number of needs are also obvious, for example in health care and the environmental protection.

A number of companies are also starting new productions based on existing demand. The active coal company (cited above) has, as one of the elements in a conversion strategy, initiated the production of consumer goods - deodorants, shampoo, knives and forks. During the first year, the company produced two million bottles of shampoo and an equal quantity of deodorants. The technical director of the enterprise pointed out that they could have produced more if they had the material resources. He also underlined that, under normal conditions, it might not have been possible for the company to sell knives and forks since specialists in this field could have produced better products.

Another company, which works mainly with aircraft engine and helicopter gear boxes, also produced tractors for the civil market. This company started producing mini-tractors, cultivators and engine units for agriculture because they expected an increase in the sales of farming equipment. To reach to the farmers, they established 18 shops in different communities around Perm and were looking for foreign contacts to establish a production of small windmills for family farming. Expectation of both increased leisure and the emergence of groups with spending power had caused a few of the interviewed companies to discuss the production of luxury yachts.

There is an already existing market for raw materials and semi-manufacture. A few of the interviewed enterprises were marketing semi-manufacture, such as, metal castings, sheets and bars since military enterprises previously had priority access to these resources. (This should continue even if they have to maintain the raw materials for mobilisation capabilities.) One of these companies was even interested in acquiring new technology for the production of more precise titanium castings and cutting instruments.

The demand-oriented conversion effort in Perm faced two major problems. First, military enterprises, which began producing consumer goods, faced competition from foreign products. As an economist interviewed in Moscow pointed out, "all those who need Russian products cannot afford them and they are only buying butter today. Those who can afford Russian products buy foreign made products."

Secondly, there is a lack of public demand. Military industries basically have a tradition of working with the public sector. Public demand would not only fit their organisational culture, but it could offer additional advantages in conversion. It could allow some high-tech to high-tech conversion - the policy of the state and the desire of the enterprises. This specific potential particularly exists in the medical field. But, there are not any available funds in the medical or environmental field to create public demand. We discovered during our interviews a number of proposals for medical and environmental applications, but they were either not developed or marketed due to the lack of funds in these sectors.

6 New Networks

A military enterprise's network was often limited in the past to a particular ministry (the dominant partner for dialogue) and other research institutes and design bureaux. Despite the fact that military enterprises, like those in Perm, are regionally concentrated, there is practically little regional interaction. Co-operation in utilisation of resources or equipment does not exist. The number of subcontractors (compared to the United States) is limited since the military enterprises are vertically integrated. In the interviews carried out in Perm, the companies were unable to point out any local partners unless they were members of the same science production association.

In terms of searching for new business networks, the Russian military enterprises face at least three problems:

- Military technology and military production are global. Production for the civilian sector is more locally and regionally rooted. Civilian production therefore implies a change of focus.
- Although there is a number of more or less serious semi-public and private consultants trying to establish themselves in this market, stable market institutions do not exist to provide marketing and distribution to civilian markets.
- The companies lack in-house capabilities for marketing and distribution.

There are clear indications that the Russian military industrial enterprises are re-orienting their networks despite these problems. Radical changes are taking place as a direct result of the mere fact that the old networks no longer exist. The League of Scientific and Industrial Associations of the USSR conducted a survey [Conversion, Rica, June 11, 1992] of the enterprises' economic ties. (The survey covered a range of enterprises (not only military ones). 82 per cent were state-owned, 7.7 per cent were turned to share holding companies, 6.3 per cent belonged to workers' collectives and 1.4 per cent were owned by local councils.)

These results show that, in mid 1992, ministerial and departmental ties already played a very small role in economic ties. The same applies to concerns and associations established on the foundations of old ministries. The aim of these associations has been to establish links to partners by using their old domestic and foreign ministerial contacts. These contacts do not seem to play a great role in establishing new economic ties. This fact was also confirmed during the Perm interviews. Most of the enterprise managers are not interested in these old, re-created structures; they prefer to make contacts on their own. Only when new foreign contacts have to be established is there a willingness among the Perm managers to use or like to use these concerns or associations.

The dominant mode of establishing networks and ties is through an independent search. This reflects the Russian managers' newly gained autonomy from a restricted, secrecy-plagued environment. It should also be noted that a third of the partners for new consumer ties "came themselves". It was quite obvious in 1992 that there was a lack of knowledge about how to find partners and, on the other hand, a great deal of activity to try and discover who produced what.

While the quoted survey includes a sample from the entire industry, the trends seem to apply to the military industries interviewed in Perm.

The enterprises are all in the process of establishing new networks. The task seems to be more difficult for high-tech production and design bureaux than for companies where both the needs and the clients are obvious (as in family farming). In cases where previous civil networks exist, the companies have a marketing lead in relation to those which had only produced for the military. There is an obvious difficulty here in approaching clients on the market. The orientation easily becomes one of technology and supplies rather than customers and markets.

7 Joint Ventures

During 1991-1992, Russian military enterprises were extremely eager to establish relations with foreign companies. One of the major strategies was to try and establish joint ventures.

In November 1991, a UN conference on "Conversion and the Environment" was held in Perm. The general atmosphere was one of "everything is possible". The West would come and help to establish - the hopes were vague at this time - a better life now that the Soviet Union was no longer the enemy. The country would no longer have to spend a dominant part of its resources on the military. The military could convert with the help of western companies. The Russians stressed, however, that they should be considered as partners not just as receivers of help.

In March 1992, while interviewing the managers about their experiences up until then, the hopes were focused on western investment and technology. Companies had drawn up very technical proposals which usually lacked any market information or information about the expected investments. The proposal often consisted of technical characteristics and a sum total of the required investments. Others enterprises were merely eager to establish contacts with foreign companies without actually having any concrete proposals. As one of the main aerospace science production concerns in the Perm area pointed out, "we should have many proposals, but we don't." Western technology was seen as a solution at a time when the West only provided Russia with humanitarian aid (1992). In spite of a number of promises about international financing, funds from the West were not forthcoming.

There is a certain paradox attached to this search for technology in the West. In interviews and discussions with Russian scientists and engineers on concrete projects for co-operation, the Russian side is always eager to underline that their technology is "world level". At the same time, they acknowledge that western technology is more advanced and is needed for their business opportunities.

Compared with St. Petersburg and Moscow, Perm is not in the main stream of foreign visitors; their level of information, therefore, is more limited. It should also be kept in mind that Perm was a closed city until only a few years ago. A company's internal KGB departments were traditionally the ones who learned about foreign achievements in technology. Foreign delegations visited Perm by 1991. A delegation from Japan visited a number of the enterprises; they were there, however, not to invest, but to apply the Japanese post World War Two experiences to the Russian situation. Delegations from American banks, American military enterprises and a Finnish industrial delegation also visited by December 1991.

During our second interview round in November 1992, the situation had radically changed. The focus was no longer on conversion since foreign or national funds were not forthcoming. However, the companies were still interested in developing civilian production and had worked out proposals to do so. This time, the search was not for western investment or technology, but for western marketing assistance. They needed marketing assistance, especially assistance to enter western markets. Many of the companies were eager to learn about how to establish a business plan. What kind of information was necessary and how can one be developed?

At this time, co-operation had been established with the Danish region Frederiksborg Amt in Northern Zealand. The regional administrations of Perm and Frederiksborg Amt frequently visited each other. A cultural exchange had also been initiated between teachers, school children, artists and others. Today, co-operation includes industrial co-operation; this led to a visit from a Perm delegation of military industries in February 1993. Although the Russian managers were still willing to establish joint ventures, they learned at this time that the western companies do not open their doors and are not immediately open for co-operation.

One of the more dramatic experiences in this direction was the Russian military enterprises exhibition, "Conversion 93", in Birmingham in May 1993. The Russian state invested a lot of resources and prestige into the exhibition; only journalists, peace activists and defence officials attended. The Russian participants, who showed their latest conversion products, were rightly disappointed. The following comments resulted from a parallel OECD seminar on conversion:

"I cannot understand it. I have a saw that can saw almost anything. I want to sell this saw, but only defence people and journalists come to the exhibition. I do not need investments; I have products to sell. I want you to buy our products. Please come and buy my products. I have received the conversion credits from Mr. Salo and Mr. Telnov. I have no contact with the

Ministry of Defence. Please come and see my products" [A director of an enterprise].

On the other hand, more aggressive tones, which underlining Russian high-tech capabilities, could be found:

"You underestimate us! Do not send us biscuits. We will enter the world market as a high-tech nation. We shall come into the market independent of what the western partners think. We don't want to start with simple little things (an American suggestion at the conference)". [A representative of an aviation firm].

The exhibition was an example of the Russian way of thinking: a desire to establish themselves as a high-tech nation on the world market. This is based on the fact that the technology push is still alive, that the products proposed were based more on technological capabilities than on market challenges and that the military is still in charge of marketing civilian products. The relationship towards co-operation with the West is becoming more frustrated, pleading or even aggressive.

One of the fundamental problems is that enterprise managers lack an idea of what constitutes a competitor. When looking for innovation they either look at their own technological capabilities or (on a very broad or general level) at the obvious needs of the market. They often have only rudimentary knowledge of other Russian producers of the same thing and how the two products compare. Many of the interviewed managers looked enviously upon their own clients and, assuming a higher profitability, were planning to begin competing production.

When looking for western partners and visiting the West, Russian managers expect to met by open doors and equal openness as they currently offer themselves to western visitors. They are astonished when the potential western partners are reserved, do not immediately see the potential for co-operation and are not willing to make on-the-spot investment decisions. If competitive advantages are analysed and it turns out that the Russian product proposal is both superior in technical qualities and can be produced for a much lesser price, the Russian managers expect immediate interest and success. It is difficult for them to understand that co-operation is a question of mutual, long-term confidence and that there might be other reasons, such as, technical performance and price, which promote or inhibit co-operation.

8 Technology Push and Demand Pull

Military conversion and the changes taking place in the Russian economy during the past few years have created a need for a new trajectory for Russian military enterprises. Enterprises in Perm have tried to achieve this by using the dual-use strategy, by bridging the technology gap, by continuing the efforts of planned conversion from the Gorbachev era and by looking for demand and new networks.

The following figure emerges if one looks at the Russian search for innovation (1) in terms of incremental and radical innovations and (2) whether the innovations are market-led or technology-driven [Fig. 1]. Most innovations are incremental and are either technology pushed - based on the existing technological possibilities - or they are market-led, which often results in consumer products with little or no value on the world market. Radical innovations, where the military's technological capabilities meet the challenges of the future, are non-existent. Hopes are attached to radical innovation, a prospect which is based on the existing high-tech base of the military enterprises. Even though some successful joint ventures have been established with the West, very few examples exist up till now. In spite of this, the hopes for integrating Russia into the world market are attached to this type of approach. Radical market-oriented innovation, which identifies industrial and social needs and provides technical and organisational solutions, exists in only very few cases. The lack of public demand, particularly in fields, such as, health, housing or environment, is the main barrier here.

Another picture emerges if one looks upon this according to another set of dimensions, namely, (1) whether or not converted products satisfy new demand or replace old products (processes) and (2) whether the produced goods are aimed at consumption or at its investment (production) [Fig. 2].

Most of the products to emerge from conversion efforts in Perm enterprises fall in this first category. These products are consumption goods which basically satisfy a new demand. Market expansion is taking place in household appliances, TV-sets, video equipment and computers. The problem is that this is also the sector where western products are competitive. Those who can afford it buy western products. Those unable to afford western products are also unable to buy Russian produced goods as well. Examples also exist of converted products replacing old products and processes, particularly in the field of cosmetics and household goods.

Fig 1: Structuring Innovation in Conversion

	Technology driven	Market led
Incremental innovation	most examples from the enterprises	a few successful examples in consumer products

| Radical innovation | the Russian conversion policy (expectation) | (environmental problems? housing? etc.) |

With investment goods, practically all initiatives fall into the category of replacing all products/processes. Perm enterprises produce equipment for oil drilling and propose to make high-strength piping for the chemical industry or machinery for the production of artificial surface materials. Very few examples of investment goods satisfying a new demand can be identified. A possible exception is filters for the chemical industry; however, they had a hard time finding markets due to lack of financing available for investments in environmental protection.

Fig. 2: Examples of Product Ideas from Perm enterprises (Structured after Edquist, 1993)

	Satisfies new demand	Replaces old product
Consumption goods	- household appliances - water purification filters	- knives and forks - shampoo, deodoran etc.
Investment goods	- production equipment for artificial surface materials	- high resistant pipin

9 Conclusion: The Need for New Policy Instruments

The military enterprises' competitive technology assets on the supply side have not been able to guarantee successful conversion and the transformation of military production into the civil sector. Even though great efforts have gone into establishing new networks within Russia and with foreign companies, these new networks have not been efficient or extended enough to provide the necessary complementary assets and a competitive edge. Companies tried to build new ventures on product areas and diversify internally while, at the same time, they created more independent business structures. These ventures are, however, still integrated parts of larger production enterprises or design bureaux. The transition problems of the Russian economy, the lack of public demand and the increased competition from western products are factors which make efficient conversion on the enterprise level a difficult endeavour.

One of the main barriers to efficient conversion seems to be the industry's and the nation's fixation on the high-tech capabilities of the military industries. The high-tech industrial base of the military industries is expected to become the bridge to the world economy. The interest in maintaining the structure of military production, i.e. large production units with high degree of vertical integration, adds to the inflexibility of resources within the enterprises and between enterprises. The interest in maintaining the existing structures is understandable when seen from the point of view of risking social unrest and large scale unemployment; but, it is detrimental from the point of view of conversion to civilian production.

The creation of new demand-oriented ventures, even in areas where product ideas exist such as environmental protection, is hampered by the lack of financing. In 1992-93, the banks in Perm would not finance conversion since their priorities were on short term projects. Regional financing for conversion was unavailable and national funds were not forthcoming. Most conversion support consisted of (and still consists of) credits to military industries for paying salaries and maintaining infrastructure. These funds could have been used instead to create an institutional structure for venture capital and/or to create a public demand in sectors where potential high technology R&D results are available - medical science and environmental protection.

Military conversion in the Perm area has been additionally hampered by the decline of demand within civil aviation. The Perm region could potentially create a cluster of innovation within aerospace and aviation. The lack of market potential has been of crucial importance for conversion efforts, military-production associations and design bureaux. Also, the energy sector could potentially be a dynamic core in the Perm region due to the oil production facilities of the Permneftorgsyntese. Although the energy and aerospace clusters have been identified as national priorities in industrial policy, the direct consequences of this prioritising process has not been witnessed in the Perm region.

New clusters could potentially emerge in the Perm region - environmental protection (the area is environmentally hazardous particularly in the area of Permneftorgsyntese) and arms destruction (where at least one company is already actively engaged). It has yet to be determined whether these clusters can develop enough in the future so as to give synergy effects and new assets to Perm enterprises. The search for advantages has only begun. The complementary assets which may become activated in the future are, firstly, the regions and the enterprises' frequent contacts with China and, secondly, the friendship offered to western visitors by Russians. Its openness and sincerity reminds us of something we have lost.

Per Wedlin

A Pskov Electronic Factory in Search of New Customers

The electronic factory Apparaturi Dalney Svyazi (ADS) in Pskov produces wire communication equipment. About 75 per cent of their products were previously used for military purposes. According the management, these orders have dropped sharply (80 per cent). The company has been chosen as a pilot project by the Swedish Network of Engineers and Scientists for Conversion (SWENESCO). Swenesco's intention is to assist the company by helping it to develop civilian products and markets and by continuously spreading information about the company's development. The purpose is to create examples of measures and results that can be applicable to other converting enterprises in the CIS.

The first step in this project was to assign two engineering students, Mathias Granqvist and Claes Östh, from the Department of Industrial Engineering and Management at Linköpings Institute of Technology to the project. Their task was to make an inventory of the company's resources and organisation for their Master's theses. Their work also included recommendations on strategies for ADS' transition to the market economy.

1 Background: Swenesco - the Swedish INESCO

The International Network of Engineers and Scientists for Conversion (INESCO) was founded in Berlin in 1991. Swenesco was founded in 1992 as the Swedish partner. Swenesco is a non-profit, practically-oriented society and is engaged in two pilot projects. The first involves the former naval shipyard No. 7 in Tallinn (now Tallinna Meretehas). The members of the society are mainly individuals but corporate bodies such as, university departments or commercial companies, are also members.

2 Swenesco's Ongoing Activities and Plans for ADS

3 ADS Strategies and Organisation

The master thesis is now being translated into Russian for distribution to ADS management. It will be discussed by the management and by experts appointed by Swenesco in co-operation with Linköping University. These experts will visit ADS for this purpose. In Spring 1994, Swenesco, Linköping University and a local partner in St. Petersburg plans to organise a seminar about strategies and project organisation. The local partner will preliminary be INGECON, a school for management of technology and economics.

The master thesis about ADS and an earlier master thesis about the naval shipyard in Tallinn will serve as a case study and starting point. Lecturers will be experts on project organisation from Sweden. The participants will mainly consists of representatives from Russian companies which are undergoing conversion from the military to civilian field. Simultaneous translation of English and Russian will be arranged so that the meeting will not be restricted to Russian decision makers with a good knowledge of the English language. The aim of the seminar is to give the participants a tool for evaluation of strategies and a brief glimpse on how a strategy can be implemented with project organisation.

4 Export Promotion

There are many export barriers to western markets in Russia such as, a confusion of languages, a lack of automatic international telephone lines, the low quality of design and manufacturing of industrial goods, a lack of price information and customer expectations, a lack of customer confidence in Russian suppliers and goods, a lack of knowledge of international trading terms and techniques, the slow transportation and low reliability of transportation methods and the fluctuating export duties and other legal hindrances.

These examples can easily be extended and is one reason why we don't believe that ADS or most other CIS industrial enterprises can export themselves out of their present crisis. The other reason is that the demand in Russia for goods that ADS can produce is significant although not yet effective. By the time purchase power rises for ADS' potential customers, the costs for expanding on the Russian market will be less than for an export effort. Enterprises with foreign owners or heavy interests may be an exception.

Russian manufacturer's present products are seldom competitive in the world market. This is also the case for ADS. However, the exchange of goods and knowledge with western companies is very important for CIS companies during the transition to a market economy. This situation (like it or not) is also a result of transferring the economic system from the west. Export orders to CIS companies can function as catalysts in this process. Export provides hard currency, enables import of components, gives practical training in trading techniques and creates opportunities for managers to visit customer companies.

Swenesco therefore tries to activate the commercial forces in Sweden. We have made advertising leaflets in Swedish for presenting ADS' resources - mainly about their plate-work capacities. As a result, a representative from a Swedish computer importer has visited ADS and negotiated a deal for manufacturing in medium volumes cases and power units for personal computers.

Three managers from ADS have also been invited to Stockholm for visits to a number of companies which have shown an interest in laying out subcontracting work. One of the companies is interested in ADS as a manufacturer for the Russian market. If this succeeds, ADS will make mechanical parts and will cover the high proportion of labour costs for the systems so that the systems can be sold on the Russian market and possibly in export markets.

There is a tendency among ADS managers to see Swenesco as someone who can supply them with western orders, i.e. as replacing the ministry which previously served in this function. Despite this fact, Swenesco will not go into commercial relations and will only act as a match-maker and promoter. As such, we make demands on them to understand that the main work must be done by themselves if they want to receive western orders or joint ventures.

In any business negotiations that requires Swenesco's involvement, Swenesco will represent the interests of ADS.

5 Communication Equipment for the CIS Market

Swenesco got the Royal Institute of Technology in Stockholm to assess the technology of ADS' communication equipment with regards to its performance and compatibility with western standards. Contacts with a technical department of a western oil company which is interested in investing in Russian gas distribution systems have also been established. The aim, in this case, is to present ADS as a possible sub-supplier for possible future projects in Russia. This is a long term prospect. The purpose of establishing this contact is also to give ADS information about communication systems for western pipelines. This information can serve as an input for developing ADS' present pipeline communication system.

6 Adopted Company Programme

To make an exchange easier between Russian and Swedish companies, a programme for sister companies has been discussed by Swenesco and representatives of Linköping University in Östergötland county. The idea is to connect Linköping University, small and medium-sized enterprises in

Östergötland and Russian enterprises. The university's role will be to support training and research. Possible sources of funds are the regional development fund in Östergötland and the Ministry of Foreign Affairs. The relations between the enterprises should basically be commercial but can also be encouraged by subsidised human software support.

7 Some Ideas from Parts of the Programme

• Giving Russian company employees some experience of working in Swedish companies.

• Providing language training in Russian, English and Swedish.

• Developing and holding regional trade fairs.

• Providing an exchange tour for Russian business people to visit enterprises in Östergötland and for Swedish people to respectively visit a region in Russia (Pskov).

• Developing courses in basic trading techniques for Russians.

Mathias Granqvist and Claes Östh

Managing the Conversion of Apparaturi Dalney Svyazi in Pskov

1 Abstract

A master thesis was conducted at the request of the Swedish non-profit organisation Swenesco, an organisation which has accepted to assist *Pskovsky Zavod Apparaturi Dalney Svyazi* (ADS/Pskov Enterprise of Long Range Communication) in its conversion to civilian production. The thesis aims, as the first phase of the conversion project, to give a description and analysis of the company's current situation.

ADS plant is in Pskov. It mainly produces communication equipment for the Russian Defence Ministry. The political and economic changes that have taken place in Russia caused a decline in military orders and, therefore, forced the company into diversification. Adaptation to a market economy and conversion to civilian production is necessary to survive in the new environment.

ADS' main strengths are its high access to qualified technicians and workers as well as its comparatively high quality standards. It also has large machinery resources in various fields of production. The analysis of the current situation indicates that a number of measures should be considered to improve ADS' position. The most urgent are:

- establishing a consensus around a clear strategy. Our suggestion is to emphasise civilian communication equipment as the main product for the future, while using non-communication products as temporary products to bridge the present decline in military orders.
- initiating a reorganisation of the company. This should include slimming down the organisation and should strive for a more flexible structure.
- tightening ties to its customers and suppliers.
- expanding the marketing unit, increasing communication with customers and investigating customer preferences more thoroughly.

2 The Study

Pskovsky Zavod Apparaturi Dalney Svyazi (ADS) has about 1,900 employees and its main product is cable communication equipment. ADS has been chosen as Swenesco's first conversion project in the Russian Federation; this study constitutes the first phase of that project. This description and analysis will serve as a basis for further work and development of the company. The report will also be used by Swenesco as a presentation of ADS to potential partners in the West.

3 The Company

The company was founded in 1958 and was mainly focused on manufacturing multi-channel cable communication systems for telephones but it also later focused on other communication systems. ADS increased its military production as the demand increased. The company has had no activities outside the former Soviet Union.

As the Soviet Union disintegrated, military orders declined and companies gained more independence, it became interesting for ADS to convert from military to civilian production. The radically changed market conditions in Russia are the obvious reason for conversion. The orders that still come from the military constitute 60 per cent of production volume but they offer only a limited contribution to the company's profit since these prices are fixed. With regards to civilian products, the company is free to set its own prices. ADS wants to extend its civilian production and hopes it will reach a 60-70 per cent share of total production soon. The main obstacle blocking an increase in civilian production is the lack of capital to invest in designing and marketing the products.

The technology used in ADS' military communication equipment is now being used to develop a communication system for a Russian oil and gas company. This order already constitutes the largest share of the civilian part of production. Civilian production (40 per cent of total turnover) also comprises products unrelated to ADS' core competence of communication technology. These products include: fruit drying cabinets, outdoor wall lamps, autoclaves for medical use and toys. Examples of products that ADS has developed or is developing on its own include: a thermostat, a blood pressure meter, a multi-voltage adapter and a bank-box system with an electronic lock. When diversification first began, ADS introduced products with a rather low technical substance; the products presently under development have changed towards ones with a certain degree of complexity. ADS is putting much hope in the communication system and the bank-box system since oil and gas companies and banks are comparatively stable.

4 Privatisation

Under what is supposed to be the world's most extensive privatisation programme, 25 per cent of ADS shares will be given to the employees and senior management - depending on how long they have been with the company. Ten per cent of the shares will be sold to employees on favourable terms (at a reduced price) while five per cent will be sold to the senior management (without a price reduction). The remaining shares are supposed to be sold at market prices. Although a further decline in military orders is expected, ADS, when agreeing on privatisation, had to ensure that it would maintain the military production line.

Privatisation will not raise any additional capital for the company since the shares are to be paid for with vouchers which were distributed to Russian citizens last year. President Yeltsin made it a special point that the shares should be distributed at no cost to the recipients. By privatising industry, the Russian government is spared the trouble of planning and financing the badly needed restructuring of these companies.

5 The Lack of Strategy

It was difficult to foresee the sudden need to convert to civilian production. As military orders were drastically reduced, it became apparent to the company that a change of activities must take place. The obvious choice was to produce for civilian markets. However, the management (except when forced to) has not paid attention to the strategic problems facing the company. A variety of products were introduced often without any common factors. After working under very stable conditions (more or less since the company was founded) there was neither strategic nor organisational readiness for the drastically changed market conditions.

The type of diversification that followed the sharp decline in military orders is not a result of an explicit strategy. New products have been taken in to utilise free production capacity. It is uncertain where the company is now heading. A hierarchical organisation lets the senior management handle a large share of the day-to-day decisions thus making it difficult to delegate responsibility and to find time to deal with strategic problems. ADS has fought with financial difficulties and has worked to survive in a turbulent environment since ties to Moscow were loosened. This is one explanation why the senior management's efforts have concentrated on operating problems. Another reason might be an unawareness of the need to address strategic problems at all. In this context we want to emphasise the senior management's responsibility to initiate the strategic work since these issues do usually not claim attention automatically. A senior management without a joint strategy and without cohesion could result in the work consequently not being conducted. The absence of objectives also makes it difficult to follow-up and evaluate the activities of the company.

6 Product Strategy

The new, non-communication products could be seen as bridging products in order to help the company survive over a period of time with decreasing orders from the military, i.e. the products bridging the gap between decline and future growth (figure 1). If ADS continues to develop these types of simple products, there is a risk of loosing competence; but, this trend is changing. Two products under development are radio communications equipment and electronic devices with a certain level of complexity. It is important for ADS to identify products which can be seen as bridging products and those which can be seen as lasting products - products to concentrate on in the long run.

Figure 1: Bridging the Gap Between Decline and Growth

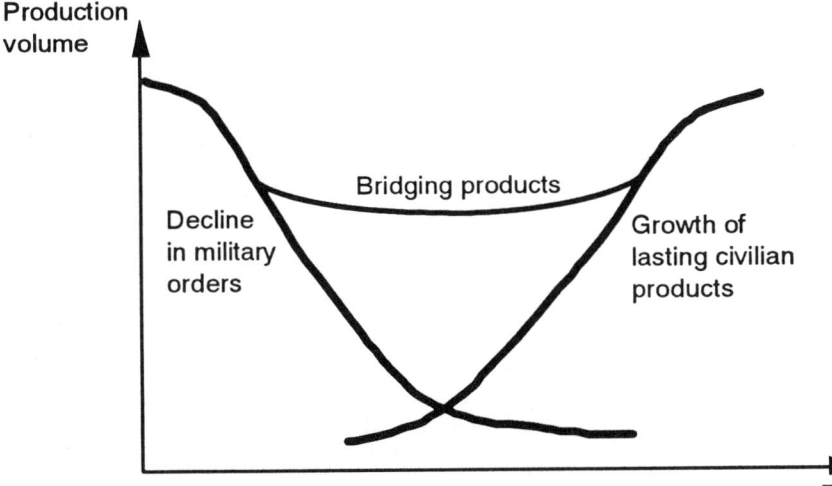

7 Diversification

ADS has already partly diversified its production into unfamiliar fields; however, the direction in which the company is now heading in uncertain. A strategy should be established before organisational changes are made.

ADS has no alternative other than diversification. Developing or changing the products for the same customer (the military) was not even considered, neither was it a realistic alternative to find new customers for the same old communication systems. For ADS, the only realistic alternative besides diversification would have been to cut production and shrink the organisation. Yet, the company has a strong sense of responsibility toward its employees and the community in general; therefore, diversification was necessary to provide employees with work.

The civilian part of ADS has diversified in two directions. There is a clear difference between products in the communication field and the other products. The recently developed non-communication products are examples of a diversification with little or no synergy in marketing or technology. One could argue, however, although the company's core competence is manufacturing communication equipment, its whole processing system - from raw is good material processing to electronics and system assembling. This means, ADS has not been narrowly specialising in communication technology. Even if there is no obvious technology synergy for these new products, the company still has the capacity to obtain this knowledge. The disadvantage of continuing this path is that the plant is not suited for mass production. The lack of marketing experience in this field is not as significant since marketing is a function that has to be developed whether it is for well-known products or for new products.

The communication system to be delivered to the oil and gas company is the result of diversification with an emphasis on utilising well-known communication technology for a civilian product. This is a new type of customer for ADS; but, some valuable experience has been gained as a result of this order and it can be used for future customer relations. Communication systems for civilian purposes is a field that should be further investigated since it has an obvious potential to become a main product in the future. The company strategy must include a means to bridge the gap between the decline in present production and the growth of future activities since the transition of ADS might take some time. In this context ADS' non-communication products should be further exploited even though they are not in line with the core competence. This is ADS' current situation.

Diversification in stable market economies is often unsuccessful, at least in the short run, due to the lack of skills and experience in new markets and technologies. Compared to advanced market economies, there are many fields in Russia where strong competition has not yet developed. It could be compared with the post-war situation in Western Europe where several markets suddenly opened for exploitation. The situation allows for diversification possibilities that would have been advised against in the West due to fiercer competition and higher market-entering barriers. In the case of Russian companies, it could be seen as a sign of inventiveness and creativity when they manage to exploit new products. The crucial issue is not necessarily advanced marketing of the product, but knowing the market well enough to produce products with a proven demand and to respond quickly to changed market demands.

8 Future Strategy and Bridging Products

Although we have concluded that there are some disadvantages with the non-communication products, they do have a purpose to fill in today's situation, namely as bridging products during the transition of the company. The products under development that are expected to have the largest capacity to succeed on the Russian market are, of course, those that should be emphasised. However, it is possible that these products will just be temporary products for ADS. It is not what ADS' plant is originally suited for - mass production of non-communication products - so competitors with less expensive mass-production costs might emerge in the future. In today's market situation, when several empty markets are open for exploitation, being first to the market is most important. Competitors will certainly appear in the future; but, by then, ADS should have managed to launch new projects in the civilian communication field, a field where there is a larger potential to develop lasting products and to be competitive in the long run. One possibility could be that one or more of the new products will turn out to be very successful and investments could be made solely to improve this production.

Bridging products could become unprofitable or could be pushed out of production by lasting products (main products) that need more production capacity. Yet, there is still a potential to utilise the gained knowledge for other purposes. It might be possible to sell ADS' knowledge about manufacturing the particular products if it can be established which plants are better suited for production of these products,. ADS would then have designs, production techniques and their experience to offer other companies. Even if the production of some products cannot be sustained at ADS in the long run, the production experience can be marketed as a service to other plants. This type of service could include drawings, production plans and even the ADS staff members who are experienced foremen that can help other plants implement their production. These services are "products" that could be able to contribute to ADS' profit. The service sector has not developed in Russia as it has in the West where it represents the fastest growing part of most economies. The service sector in Russia will probably expand in the future.

Bridging products and lasting products must clearly be identified. Bridging products must have the capacity to generate a positive cash flow for a period of time so that the company can survive until lasting products fill most of the production capacity. Bridging products should not be allowed to compete for major company resources since their production should stop when capacity is needed for the lasting products. The company strategy determines what range of products will be the main products; but, of course, the strategy can (or rather should) be regularly revised in order to adapt to changes within and outside the company.

9 Organisation

As in most companies in the country, ADS' organisation could be described as mechanistic, i.e. it has a rather hierarchical structure where different units are highly specialised and fields of responsibility are well defined. Companies are used to acting in a very stable environment; there was little incentive to develop flexibility or even preparedness to adjust to altering conditions. Strategic decisions were made in Moscow and expressed in long term plans which had to be followed.

Today, five-year plans have been abandoned and companies have been forced to start their own planning. In market economies, the market environment tends to fluctuate and companies must be prepared to make adjustments to handle these sudden changes. This is especially true for the turbulent situation in Russia today. The economic and political situation is quite unpredictable and, therefore, signifies the need for more flexible organisation.

Quick decisions, either to parry the changes or to adapt oneself to the changes, are needed. In order to quickly respond to changes, warning signals must be detected as early as possible while adaptations of the organisation must be facilitated so as to prevent negative consequences. A flexible organisation, which is capable of adapting itself to new situations, requires individuals with an open understanding of the necessity to adapt themselves and the organisation whenever change occurs. Key persons must be vigilant and must be able to detect warning signals; moreover, the organisation must be permeated with the aspiration for a rapid response to changing circumstances. Extensive organisational changes are certainly difficult to accomplish rapidly, but we suggest that the management try to attain this kind of more organic structure during the continuous effort to re-organise ADS.

Flexibility is hampered by an unusually high share of white-collar workers and makes the organisation more bureaucratic than necessary. Some 600 employees have left the company during the last two years; most workers were dissatisfied with their salaries. This left ADS with a 40 per cent share of white-collar workers. Slimming the organisation would cause the decision-making process to speed up and reduce costs for underemployed administrators. However, the problem of what to do with the redundant white-collar workers remains. Many employees would probably have to accept degradation if alternative occupations are to be found for them. Nevertheless, measures still have to be taken to trim the organisation.

Further training and education is an alternative if the company does not wish to lay off personnel. This would broaden knowledge to span over new fields and would be a part of the foundation for the new organisation. It does involve some costs; but, this should be seen as an investment and is much more preferable to letting employees with a salary stay unoccupied at home. Even if many employees are highly skilled today, their skills are often in a rather narrow field. Further training can make the individual capable of performing various tasks and this flexibility is the lubricant of a dynamic organisation. It would keep presently underemployed employees active and make them better prepared for various tasks. To maintain the company's competitiveness the strength skills and knowledge of the employees should be continuously improved. Any re-education must be designed according to an established company strategy and should take into account the human resources the company can make best use of to reach its objectives.

In companies with unit production such as ADS with its communication systems, each product is customised. It could be an advantage in these companies to organise the activities according to the projects that are offered and sold - project organisation. Project organisation, which is created for a single order or customer, makes it easier to include the views of the customer since the project leader can work closely with the customer. It is also easier for a customer to identify the right person to contact and to have a feeling of receiving personal treatment from the company. This is already useful in the production development work both to ensure that the customer really gets what he asks for and to ensure that close contact with the market in general is maintained so that early signals about what the market and the customers demand can be picked up.

The above discussion implies that there are several reasons for getting underway with the restructuring of ADS' organisation as soon as possible. Flexibility is needed for a quicker response to customer demands in this rapidly changing environment. This would increase ADS' competitiveness. Measures to slim down the organisation, to reduce costs for prevalent overlapping work and to stop the underemployment of employees will also increase flexibility.

Ksenia Gonchar

State Industrial Policy in the Defence Complex

The events of October 1993 in Moscow have, without a doubt, created a rather new environment for conversion policy, have provided additional unfavourable circumstances and have stimulated a considerable change in the institutional and legislative context of defence economic restructuring. Although the documents illustrating theses changes have appeared right before our workshop and although the outcome of the political games is very unclear, it seems reasonable to present some preliminary and disputable assumptions. It might be suggested that the military and the higher leadership of the defence industry are experiencing a kind of a political rehabilitation with the last chance, given by unexpected loyalty during the coup, to exert pressure on the government. Such pressure will not necessarily be in favour of militarisation, but probably with the aim to avoid the common regular measures of structural and financial policy, to gain effective control over the defence industry and to save the most valuable enterprises from the further technological degradation and market uncertainties.

The general economic situation does not allow for more reliable forecasting. The fourth year of economic decline has resulted in a 38 per cent decrease in GNP (1990-1993), a 47 per cent decline in industrial production and a severe investment crisis. (In 1993, the share of investment in GNP accounts for only nine per cent with inflation in the investment sphere at a rate which is higher than even within the wholesale or consumption market - 38-44 per cent per month as compared to 23-26 per cent - and an extremely low marginal efficiency of federal budget investment - seven kopecks on one rouble. Budget credits and subsidies are mainly used for current consumption or they are frozen in incomplete construction.) These figures do not even take into account the negative rate of capital accumulation by nearly all machine-building industries. The new Russian paradox is that, in 1993, the unemployment rate has decreased in comparison with the level for 1992 - it accounts for one per cent.

Under these circumstances, the country is unable to afford an urgent state industrial policy which is aimed at prolonging any further decline. (Even the most devoted and romantic adherents of liberal economics call for such measures.) We can identify at least five different and mutually exclusive industrial policy concepts which compete as policies for political lobbying:

1. the government's (or more correctly, a team of liberal Vice-Prime Ministers') concept;
2. the Defence Ministry's concept;
3. the concept from the State Committee for Defence Industry and the State Committee for Industrial Policy;
4. the independent line taken by a powerful group of defence enterprise managers who are united by the League of Defence Enterprises;
5. and the exotic actions of the President himself - they can hardly be defined as industrial policy, but, as a matter of fact, do have a strong influence on it.

Although this characterization is a bit artificial, every line has its own supporters outside the main political and strongly scattered groupings. The development of new parties and alliances on the eve of elections also adds to the variety of views on state industrial and defence industrial policy. But, it should be stressed that most of the concepts are nothing more than plans or intentions while the newly formulated decrees present a kind of compromise in which unidentified interest groups can directly benefit from their implementation. So, if these documents are applied to a particular line of industrial policy, it is more an aspiration for finding any logical order than a strict belief in any certain system.

The differences between the above mentioned lines can be discovered by understanding the nature and instruments of the structural policy, federal and regional programmes and institutional framework - mainly in the context of privatisation and its shaping of the so called financial-industrial corporations. The following is an analysis of the two main lines of state industrial policy.

1 The Government's Liberal Line

The government insists on the tightening the financial stabilisation policy (with the aim of decreasing inflation to a level of 15 per cent in January 1994 and to a level of three per cent in December 1994), sharply lowering investments (from a level of nine per cent of the GNP in 1993) and providing credits and the conversion of debts into freely circulating exchange bills. This structural policy was described in the latest Statement of the Council of Ministers, Government of the Russian Federation - "To the Economic Policy at the End of 1993 and 1994" (issued in November 1993) - and suggests that the following will take place:

- a repeal of export quotes on non-fuel goods and export taxes for the majority of products (It could lead to the growing export of capital because the liberalisation of the export policy is combined with restrictions on currency exchanges in the domestic market.);
- a cancellation of subsidies on imported goods (Consumer durables and civilian high-tech production in the defence complex use up to 30 per cent of imported components. If this measure is accompanied by a high import tax on "the goods of the elite group and products of non-urgent need" then prices will stand still or increase far above the world level.);
- an elimination of all land property limitations and real estate trade (Such rules are advantageous for defence R&D and producing enterprises with a low dependence on defence orders because they can benefit from high land prices (especially in Moscow and St. Petersburg) and can include land and premises manipulations into a strategy of managerial behaviour.);
- an elimination of profit limitations (This is a good chance for "permitted" monopolies.);
- an immediate implementation of the law on bankruptcies.

As we may understand from this strange document, the main instruments of industrial policy will concentrate on liberalising foreign trade and on creating a structural role for bankruptcies in loss-making enterprises.

As to the mechanisms of financing of state investment, it is suggested that budget money will be distributed only in the form of investment credits and will be used only for attracting the further flow of private capital as an independent means of supporting certain branches and enterprises. Moreover, budget allocations will be given in exchange for shares according to the "state money-for-equity" scheme. A small part of the state investment will be given to the investment foundation with the aim of providing insurance for private investment in the sphere of national interest. The President of the Russian Financial Corporation, A. Nechaev, as a matter of fact, has admitted that financial stabilisation is a priority over structural policy and that their intentions are to finance projects with only a very short-term return of capital (no more than two years). Under these circumstances, the future of federal programming - the majority of programmes call for trillions in budget expenditures - looks very unclear; the role of government experts will most probably be aimed at killing these programmes.

The institutional aspects of governmental plans which concern industrial policy may be traced back to the new State Programme of Privatisation (to be adopted in November 1993). Although the text of this document is not yet available, it is known that:

- the programme increases the number of enterprises prohibited from privatisation;

- it specifies the procedure for privatising enterprises from different branches (including commercial banks and establishments already undertaking conversion);

- it gives priority to privatised enterprises in comparison with state-owned firms (excluding those in which privatisation is banned) so that private enterprises can receive of all kinds of state support - subsidies, credits, tax advantages, access to resources and help from foreign technical aid and Western credits. Moreover, regions that are successfully implementing privatisation programmes will be encouraged by special advantages; those who fail will be punished by sanctions;

- it subordinates firms which remain as state property to state controlling bodies and it loosens any rights provided under the old Soviet laws concerning enterprises. Firms from the veto-list will be converted into "establishments financed from the state budget" and will have very little space for independent decision-making.

The point about financial-industrial corporations in the governmental version of industrial policy is very unsound and, up to now, was only publicised by M. Yuriev in the form of vague "multi-profile industrial groups" - they are understood to be an alternative to the existing privatisation strategy.

What role do defence complex enterprises play in the governmental industrial policy? The most probable answer is that they are taken out of the general scheme of industrial and economic policy and receive special status in all its aspects - from financing procedure to privatisation norms. The liberal government has given the defence establishment control over the defence industry and its civilian restructuring. The defence industry, however, will remain under the supervision of the Defence Ministry and the State Committee for Defence Industry (organisations which do not share the same views and have conflicting interests as compared with the defence industry).

Two recent Presidential decrees, "On the Peculiarities of Privatisation and Additional Measures of State Control over the Functioning of Defence Enterprises" (August 1993) and "On Stabilisation of Economic Situation of

Enterprises and Establishments of the Defence Industry and Measures to Guarantee State Defence Orders" (November 1993) reflect this suspicion. It is unclear whether this situation is good or bad for the outcome of transition reforms and conversion results. Both approaches are full of contradictions - sin against real economic trends are not supported by the defence industrial elite (their common consent) and, therefore, do not have a lot of chances for full implementation. But it still seems important to look at the mentioned documents and the defence establishment's creativity in industrial policy-making.

2 The Defence Establishment's Version of Industrial Policy

It is impossible to identify the authorship of the concepts, political lines, final decrees and decisions which reflect this line of industrial policy. There is reason to believe that these ideas were mainly influenced by a structural policy concept that was produced last year by a team headed by Deputy Defence Minister A. Kokoshin and by the so called "Programme of Oleg Soskovetz" that was presented in early November. Although these are different approaches, they seem to express the most powerful pressure on the adopted procedures of state control over the defence industry and converting enterprises. Little is known about the above mentioned programmes (they are not yet published and are not available for public discussion), but the following assumptions can be made about the key elements:

1. The decree on economically stabilising defence enterprises definitely protects them from a strengthening of the financial stabilisation policy. This notion confirms that defence industrial restructuring is exempted from the mainstream of economic policy, confirms that the criteria for economic efficiency cannot be applied to the defence complex and guarantees that some special privileges will continue - implementation of the cash-in-advance system of progressive payments for state contracts, special rights to include labour costs equal to eight minimal salaries (or ten in the nuclear industry) into the costs of production (in non-defence industries this rule is limited to four salaries). It was previously stated that the new state investment models are not applied to the defence complex; they will receive budget means according to the traditional procedures.

2. The decree on privatising defence enterprises exempts 474 establishments out of 1,700 from the privatisation process and stops all procedures before the new veto-list is produced. The privatisation of enterprises which formerly belong to the defence branch is conditioned by signing an agreement with the government that concerns a responsibility to fulfil state contracts and to keep state secrets safe. If a defence enter-

prise is being privatised, the dividends from the state portion of shares are to be invested through special financial institutions into the conversion and social support of the staff. Directors of enterprises in the defence complex (including those who have received the right to privatise) will receive a special certificate from the Council of Ministers according to the recommendation of the State Committee of the Defence Industries. (This is mainly a powerful instrument which will be used to exert pressure on unruly managers.) Maintaining the defence industrial elite under ministerial supervision is accompanied by getting rid of loss-making or obsolete enterprises so as to lessen the burden on the military budget. High-tech enterprises which do not have to produce equipment for the new military doctrine may also find themselves among the outsiders (this probably refers to missile factories).

3. The State Committee for Industrial Policy presented the "Concept and Mechanisms of Implementation of State Industrial Policy for 1992 and Middle-term Perspectives" and a separate volume of defence industrial policy prepared by the State Committee for the Defence Industries. These documents seem to be rather sound and influential, but they still contain a rather uncertain perspective. These programmes are oriented towards investing ten trillion roubles into the strictly selected support of certain enterprises and branches. The direct control over state-owned enterprises - defence, export-oriented, "dangerous" or enterprises that are economically inefficient, but vitally important to the national economy - will be replaced by financial-industrial corporations instead of branch ministries. FICs will be able to identify real bankruptcies from normal debtors and will forgive privatised company bankruptcies at the expense of having their equity returned back to state ownership (debt for nationalising model). Although these plans greatly limit the freedom of managerial behaviour, the plans could find support among directors of factories that are badly in debt. Directors, in this case, might prefer loosening their independence in the hope of receiving enough financial support to survive. Some of the "victims of conversion" - those who lose the new privileges offered to the defence complex - may also choose this type of formula.

4. The concept of financial-industrial corporations is not yet clarified and, at this time, includes a vast variety of opinions: from holdings which are a substitute the former ministries to the forced huge units of commercially efficient factories where enterprises produce weaponry and have dependent banks. Up to now, several formed groups are already based on the principle of mutual technological supplements. The idea is to consolidate general contractors with the supplying and subcontracting units

(this could include the consumers of the manufactured equipment as well). Under the threat of inevitable conversion and privatisation, the State Committee for Defence Industries started to organise new formal and informal alliances with financial institutions so that they could prevent outsiders from participating in the privatisation of defence enterprises. For example, the Committee signed two remarkable agreements in September: one concerned a specialised investment-privatisation foundation called Conversion-Hermes and its "co-operation in the field of investment activity and the shaping of the securities market in the interest of privatising enterprises and organisations of the defence industry"; the other concerned a similar treaty with the voucher investment foundation called Military-Industrial Complex.

5. Although the federal conversion programme was issued several months ago, it does not yet play a significant role except between the ministerial functionaries. This programme was mainly developed in the first half of 1993 by the Ministry for Economics and it is highly influenced by the Gosplan tradition. The programme (1993-1995) states that the main goals of conversion should be the preservation of the professional, industrial and technological potential of the defence complex, the import substitution of non-Russian production, investment into the techniques and technologies of economising energy and resources and the maximisation of utilisation capacities. It includes fourteen special purpose programmes - civilian aircraft industry, shipbuilding, equipment for fuel processing, forest industry, housing and road construction, equipment for agriculture, textile, food processing, trade, consumer durables, communication equipment, conversion for environment, medical equipment, and the programme of conversion for the enterprises of the Ministry for Nuclear Energy. The programme costs 325 billion roubles of budget subsidies and 300 billion roubles of privilege credits (in the current prices of early 1993).

The expected result is a 12 per cent growth in civilian production from defence enterprises. This strongly contradicts the expected further economic decline (forecasted by another department of the Ministry for Economics) and the sharp decrease of civilian output from defence enterprises this year. In conclusion, programming as an instrument of industrial policy in the time of emergency presents an example of wishful and unrealistic thinking which is far from the real conditions and needs of the Russian defence industry.

Martin Salamon and Ian Whitman

Conversion of the Labour Force: OECD Training Programmes in Russia

Within the Division for Education and Training, the principal activity concerning the Newly Independent States (NIS) has looked at the human resource issues of the transition process in general and the conversion of military officers to the civilian economy in particular. In 1992, the OECD's Centre for Co-operation with the Economies in Transition (CCET) carried out a study of Russian efforts to retrain the very large and increasing number of demobilised military officers. This report [GD(93)9] did not attempt to cover the conversion efforts of member countries such as the large programme carried out by the Federal Republic of Germany.

Russia possesses a great number of higher education institutions at the technical university, vocational and tertiary levels. The problem is that these institutions were geared towards preparing young people for employment in the public enterprise sector, although there are now moves to emphasise private sector themes particularly in regards to readapting the military-industrial complex to privatisation.

These institutions are organised into networks of schools and most of these networks are involved (in different extents) with the conversion of military officers and civilians employed in military industry. The most prominent institution is the Committee for Higher Education. This is because of its responsibility for accreditation and its right to grant diplomas. The Federal Employment Service, the Academy of Defence Industries working with the Foreign Ministry and the Ministry of Defence also operate training institutions with some responsibility for conversion training. All of these networks have attacked the problem of retraining personnel. Their efforts cover the majority of subjects in western macro-economic and micro-economic courses; but, these efforts generally suffer from the same shortcomings of :

- trying to cover too much material in a short time span,
- sacrificing practical information to market theory and behaviour,
- not being suitably adapted to the Russian reality,
- and competent teaching staff which is lacking in first hand knowledge of the
- workings of a market economy.

The pilot project proposed in the OECD report would address these shortcomings and, at the same time, it will:

- attempt to establish a strategic programmatic framework for co-ordinating the efforts of donor countries and institutions to achieve a comprehensive and coherent action plan,
- mobilise donor resources to prevent overlapping programmes,
- provide information on existing efforts,
- and set up a management structure to evaluate the system, monitor programme implementation and respond to changing needs.

On the practical level, the training of trainers from the above mentioned networks would be carried out at a centre located in Russia and through extensive hands-on training in public and private member country institutions and companies. These trainers would then be responsible for conversion training in their home institutes. The number of officers who complete this training could be greatly increased - from the 2,000 trained in 1992 to the 8,000 projected for 1993. At the same time, a relatively small number of selected officers would be trained in areas such as small business development, labour office management and counselling.

A recent World Bank report favours the establishment of a Russian Training Foundation (RTF) with three regional affiliates to address the problems of cross-sectoral activities in key marked oriented fields. This partnership between the government and private consumers of skills (like the OECD project for officers) would depend on existing external donor programmes and whether new donors could be found. This type of programme would have an important multiplier effect in establishing the demand-oriented training for market skills that are necessary for a successful transition.

There are close relationships between education and training on one hand, and employment on the other hand. But, these links need to be considered carefully as one cannot assume that more learning will automatically lead to new jobs. The Russian situation is unique because the work-force is highly educated and because Russia does not suffer from problems of literacy or lack of fluency in the native language as in the case of many western countries.

The success of economic transition depends on popular support; education is a key area for achieving this end. A number of specific points can be made that are equally true for the education of the future work-force and the re-skilling of adults:

1. Decentralisation is not a simple matter of the central government giving up authority or attempting to impose a western style system. It requires a change in attitudes and the development of management and decision-making skills as well as the reinstatement of democratic processes at the regional local and institutional levels.
2. The classroom must reflect new teaching and learning styles and should be adapted to more active participation by both students and adults.
3. The pace and content of change can be accelerated if educational reforms are demand led.
4. Wider use of competency-based learning systems which focus on achieving outcomes could also increase consumer choice by making it possible for institutions beyond the formal education system to provide education and training.
5. It is necessary to include all of the stakeholders.

Numerous OECD countries have experienced military conversion and, although the military industrial complex is not nearly as pervasive in these economies as in the NIS, some of their experiences of education and training could be useful for the Russian situation. Recent reports by the Massachusetts Institute of Technology indicate that the Taylorist model of production with its hierarchical structure and heavy administration may have been a cause for significant losses in the competitive position of the US. To counteract this trend, many American companies have looked to countries like Germany and Japan to learn lessons form the production, work organisation, technological integration and training methods used in their efficient business operations. In Japan, on the job training, rotation in different departments, distance learning and quality circles create not only a trained workforce, but a different type of worker. In short, it can be said that the type of work organisation that develops will have an impact on the skill structure.

As the enormous multi-production state owned companies of the NIS become privatised and change to civilian consumer production they will face the challenges of a more knowledge-intensive economy. Among the principal features and developments at enterprise level, the following can be singled out as being particularly important:

- Firstly, new technology provides new options in terms of work organisation and skill structures;

- Secondly, human resources will become a new competitive parameter for enterprise as competition grows, and, consequently, strategies will have to be developed to better co-ordinate tangible and intangible investment;
- Thirdly, there will be strong pressure to produce higher quality products and services with shorter life spans. This will lead to an increased need for the integration of new technology, work organisation and skill formation which, in turn, will require principles such as teamwork, and decentralisation of management and responsibility;
- And as a fourth point, routine and low-skilled jobs will decrease, while new jobs will demand higher and higher skills.

These general trends have been visible across enterprises and countries. Recent OECD research has encompassed a variety of human resource strategies that enterprises have pursued to cope with this situation.

The first could be called "the human resource intensive strategy". In this strategy the enterprise recruits people with a good broad-based educational background and then complements this with intensive on-the-job training and education coupled with a flexible work organisation to allow for job rotation.

The second could be called "the polarisation strategy". Enterprises adopting this strategy tend to focus their human resource development on a core group of employees through innovative measures in skill formation and work organisation. Outside this group of core employees there exists a buffer group with less formal ties to the human resources development of the enterprise; these groups are often highly vulnerable to shifts in business cycles.

The third could be called "the mobility strategy". Within this strategy enterprises tend to recruit highly educated personnel and offer little in-service training. Learning and earning is part of the same work process and inter-firm mobility is very high.

It is important to realise that these strategies and their combinations and variants depend on the education system, the labour market and the training level of the country. "The human resource intensive strategy" seems easy to implement in countries where the formal education system manages to provide a good level of general or vocational education at upper secondary level coupled with a tradition of labour market flexibility. "The polarisation strategy" is often found in countries with weak vocational education systems and a relatively high rate of failure at the secondary level coupled with an inflexible internal labour market. "The mobility strategy" often exists in countries that have a very competitive high technology and private service sector with strong labour market mobility and a responsive system of higher education. In the NIS, we will probably see variants and combinations of these strategies depending on the activity, regional conditions and future orientation of the defence industry.

The principal message that emerges from analysis of target groups and providers of education and training in relation to human resource development is the need to revitalise in policy terms the strategies for lifelong learning and recurrent education.

Peter Lock

Supporting conversion: A First Approximation[1] to an Alternative Approach

The issue: As recently as 1992 the "peace dividend" was given a prominent status in UN-reports and put at 1.2 trillion US dollar for the last decade of this century[2]. The wishful linkage between disarmament and potential development was also the major ideological pillar of Gorbachev's "global mission" that was designed to preserve the Soviet system. [3] But, even his successors who accept the need for a systemic change are faced with the reality that the intended conversion of the military-industrial complex made no or only insignificant contributions to the well-being of the Russian population. Both had overlooked the structural economic devastation the arms race had inflicted upon industrial structures world-wide.

The combination of an obvious over-militarisation of the economy with systemic characteristics of a "permanent Kriegswirtschaft (war economy)" [4] as a point of departure made successful conversion a concomitant (pre)-condition of transformation. Under these circumstances it is not surprising that conversion was at the top of most agendas that deal with the stabilisation and transformation of Russia. From the perspective of text book requirements for a smooth transformation, the performance of the acting governments has been dismal. Resources continue to be wasted on maintaining structures that will have to close down in the environment of a market econ-

---

1 The author's backgrounds includes several visits to Saint Petersburg since 1990 (the last in September 1993), recent literature on the failed conversion of the Russian military-industrial complex and extensive initerviews with western business representatives, Russian economists and Russian managers. Interviews with Russian managers were not carried out in September 1993.

2 United Nations, Human Development Report, New York, Oxford, 1992, 86.

3 Åslund, Anders, Gorbachev's Struggle for Economic Reform, London 1991

4 See Sapir, Jacques, Les fluctiations économiques en URSS 1941-1985,Paris 1989 and Sapir, Jacques, L'economie mobilisée, Paris 1990

omy, while necessary investments for modernisation are not made. Many crash programmes were suggested, few were launched, hardly any was turned into a success. However, the Russian governments have so far managed to avoid a total collapse of the economy, in spite of the gloomy predictions many economists echo as government indecision continues. History can only qualify this performance as either a total failure or a relative success.

The most well-intentioned western observers generally overlook the fact that any successful transition strategy must be based on a realistic assessment of the status quo ante; even the most radical policy will be determined by parameters of historical continuity. Seasoned within the old system the industrial leadership of the VPK managed to exploit the existing uncertainties[5]. The old networks formed a cohesive interest group which rapidly learned how to survive in the new climate of postulated transition, though not without continuing to pull strings of the former system many of which remained operational. They also managed to improve their personal situation.[6] In the absence of a social security system the government was forced to grant privileges and to legitimatise ex-post-facto substantial volumes of money created "subversively" through reciprocative inter-company credits. Their de facto function as a substitute for the non-existing social net provided the large enterprises with an efficient leverage vis-á-vis the government. However, not every enterprise[7] and research institute[8] that was closely associ-

---

[5] In the old planning system, withdrawing additional government credit above the limits set in the plan was punished rigourously with an interest rate of 30 per cent. The inertia of old rules provided the VPK a source of highly subsidised credits since the monthly inflation rate approached this percentage. (I am indebted to Dr. Ksenia Gonchar for this information.)

[6] The wage differences within the hierarchy of enterprises have considerably increased and have favoured the top layers.

[7] High-tech companies continue to lose qualified personnel due to low salaries. AW & ST reports a new wave of engineers quitting their jobs. Jeffrey Lenorovitz, Boris Rybak <u>Engineers Flee Low-Paying CIS Jobs</u>, in: Aviation Week & Space Technology, September 27, 1993, pp.53-54.

[8] Some research institutes are particularly vulnerable since the government budget for R & D continues to shrink in real terms. For a

ated to the VPK were equally well situated and connected "to muddle through" and withstand structural and monetary pressures for rapid restructuring and conversion.

In the wake of reduced procurement examples of successful industrial conversion at the plant level are scarce in western countries. Companies often respond by diversifying into new branches by means of acquisition. Military production sites tend to be closed down and "converted" into real estate rather than into civilian production. Moreover, there are startling examples were defence manufacturers have abandoned their traditional sites and moved their production to entirely new plants in order to stay competitive[9]. Against this background, it is surprising that western experts tended to assume that conversion of military production, while failing in the West, should provide an operational basis for a sound economic strategy of rapid transition to an internationally open market economy in the FSU.

Early optimism has since faded; it is doubtful as to whether an unrelenting search of the causes of the undeniable failure of conversion in the FSU has begun. Conversion does not prominently figure in the political debate in Russia any more. The general fear of social explosion and political upheaval once restructuring gains speed in a country where an institutionalised social net does not exist has contributed to a permissive climate among western observers[10]. An additional closer look[11] at the transformation in the former GDR where the social net is being provided by unrequited transfers from the

---

comprehensive overview of the decline of r & d see: OECD Science, Technology, and Innovation Policies - Federation of Russia - Background Report, Paris Sept. 1993 as well as the corresponding evaluation report.

9   Rheinmetall abandoned its traditional plant in Düsseldorf and moved production to a new industrial park in Lower Saxony. For a detailed description of a similar case in Great Britain see: James Buxton Sights Set on a New Target, Financial Times, May 12, 1993; pg.11.

10 A consensus seems mature that the transformation in Russia will have to follow a slower pace than the early monetarist approaches presumed.

11 Lutz Hoffmann, the director of the German Institute for Economic Research (DIW) provides a sober account of the lagging transformation: Warten auf den Aufschwung, eine ostdeutsche Bilanz, Berlin 1993.

former FRG is likely to suggest that entirely new approaches and time scales are imperative in order to design a viable strategy of transformation for Russia.

One of the important reasons contributing to failure and even resistance to all conversion efforts is the inherited industrial organisation. Institutional production and the innovative potential of R & D were generally separated under the old institutional set-up. The reorientation of R & D towards self-financing civilian ends appears to be additionally hampered by the institutional distance between production of civilian goods and the existing R & D potential. A survey of high-tech industries in Saint Petersburg reveals profound differences between the Russian status quo of organising R & D and corresponding structures in the West. Dynamic innovations in the civilian economy are amalgamated within the globalising economy by transnational corporations[12].

Integrating basically military-oriented research institutes into the circuits of innovation of the global economy would constitute a successful conversion of the inherited militarised R & D potential in Russia. But, apart from the immobility of these institutes who prefer to adhere to the rules of secrecy of the VPK rather than advertise their capabilities, western experience highlights the difficulties of organisations operating within the military realm to supply spin-off technologies to the civilian economy. [13] The "military culture" is generally not conducive to productive flows of knowledge towards the civilian sector[14].

---

12 For a comprehensive account of global innovation strategies see: OECD/TEP <u>Technology and the Economy, The Key Relationships</u>, Paris 1992.

13 OTA, U.S. Congress, Office of Technology Assessment, <u>Building Future Security</u>, Washington D.C., 1992.

14 A recent symposium on conversion organised to celebrate the 25th anniversary of the University of the Bundeswehr (armed forces) confirmed this observation. "Insiders" provided many illustrations of the specific enterprise culture in the armaments sector which almost exclude conversion to succeed other than direct subsidised (non-market) government procurement of civil commodities. One panelist stated that there has been no successful conversion at the plant level in Western economies, his statement was not contradicted by a description of single successful case.

A survey of the high-tech sector in Saint Petersburg reveals that the region was a focus of Soviet military-industrial efforts. The survey commissioned by the Dresdner Bank was prepared by the Central Design Bureau "Rubin". Although it warns the reader that it is not a comprehensive compilation, it lists 83 research institutes and 98 high-tech enterprises in the manufacturing sector. Many of the latter additionally claim to have their own R & D departments. Less than 100 out of 181 enterprises reveal the number of employed personnel which amounts to 350,000. 44 of a total of 83 research institutes employ more than 80,000 persons, approximately 2,000.on average The corresponding figure for high-tech manufacturing approaches 6,000 persons. To the extent the range of products or research was revealed at all, it is safe to assume that at least three quarters of the surveyed institutions formed and most likely still form part of the military-industrial sector. When comparing the quantitative dimensions of these research institutions with the way efficient civilian R & D is organised in western industrial countries, it is obvious that conversion of existing R & D infrastructure will be a particularly difficult task.

The extremely large sector of R & D institutions in Saint Petersburg was nurtured by military orders and shrouded by absurd levels of secrecy. Military orders were cut dramatically, but the culture of bureaucratic secrecy was not questioned. Torn between the faint hope that military demand would catch up again and the imperative to completely open up to international or hardly existing internal markets, a large number of managers were not prepared to meet the minimum requirements of transparency market entry requires. Almost half of the high-tech enterprises in Saint Petersburg did not declare the number of persons employed or concretely the type of activity they are involved in or both. [15]

The manufacturing sector Saint Petersburg is burdened with the fundamental weaknesses of the Russian industrial sector, namely, a vertical integration of production which does not allow for an efficient division of labour and scales of production due to specialisation. The inherited production philosophy was geared towards relative autonomy; its imprint includes outsized factories burdened with dozens of inefficient workshops and up-stream production lines. At the same time they are absolutely deficient in terms of any down-stream activity marketing, after sale service, etc.

---

15 DOWC, 1992. Deutsche Ost-West Consult St. Petersburg's High-Tech Sector: Company profiles and technical proposals, Frankfurt/Main Dresdner Bank.

If, in this context, a military plant aims at producing civilian commodities, it is bound to draw upon in-house up-stream supplies. The resulting product is likely to reflect technical features which are defined by the militarily conditioned up-stream production capacities rather than any market oriented properties. Examples of such conversion, which are bound to fail once they are exposed to market conditions, were presented in a special spring 1993 exhibition in Moscow (BDNX). They included a vacuum cleaner an average housewife is not capable of lifting and a multi-purpose agricultural tractor which does not meet basic ergonomic criteria.

It seems that for the time being a majority of military factories have entered a defensive cycle where available funds[16] are used to maintain the workforce at minimum wage levels[17] often by reducing working hours. Sometimes the production of military items continues even in the absence of government orders. The managers do not pursue a vigorous drive to produce for civilian markets. They often seek a western partner for their high-tech sections and try to hive-off these sections from the main enterprise in order to form new privatised companies[18] that are under their personal control.

Finally, a strategy of conversion presupposes a vision of the future defence industrial sector in Russia in order to make rational decisions which capacities should be maintained and modernised and which capacities are either to convert towards civilian production or to close down.

Although some time will pass before a consensus on the future defence posture of the Russian armed forces emerges and guides industrial policy, it is possible to define the basic parameters which imply profound changes in the structure of the Russian military industry. The imperatives for the development of its future defence industrial base cannot be totally different from new industrial paradigms which presently guide the restructuring of military

---

16   The deviation of investment funds supplied by the government, income derived from real estate activities and sales of existing stocks of semi-manufactures figure prominently among the sources of these funds.

17   These minimum levels are often supplemented by payments-in-kind earned through barter transactions or by the company organising wholesale purchases of consumer articles on behalf of the employees.

18   In the absence of clear legal rules, such moves require the collusion of the state bureaucracy, thus opening up gateways of corruption.

manufacturing in western countries. Since Russia is bound to abandon its splendid isolation.

The evolving industrial paradigm of future defence bases takes into account the fact that advanced military technology increasingly relies on technology and innovation created within non-military global industrial networks. As a consequence, western defence industries are under pressure to change their design philosophy and to restructure in order to be able to continuously absorb innovation[19] as it becomes available in world-wide civilian markets.

Modulisation and dual-use are the catch words of this important evolution. Originally affirmed dual-use properties of military items were instrumental to overcome export controls, particularly in Europe. The military lobby and large American military laboratories, were recently eager to usurp the term in order to legitimise the continued government support of their institutions. Dual-use actually refers to the integration of civilian innovation into complex weapon systems at ever larger scales. It is not an entirely new paradigm. However established military R & D suppliers cannot simply adopt it, so as to survive. It presupposes profound changes in the corporate culture.

It is important while the discussion is still going on, to consider this new industrial paradigm when taking decisions on the future defence industrial base in Russia. The past achievements of certain actors in the defence sector may not be a warranted indicator for orienting future defence industrial policies which must have the potential for gearing the sector towards the challenges of the 21st century. If it is truly seen as responding to the challenges of moulding the sector into the globally evolving industrial network, the new set-up of the defence industry is likely to diverge considerably from the present hierarchy and sector composition.

Structural changes within the world market will orient Russian strategies of transformation and conversion, particularly in the defence sector are past experiences of little value. Only a cost-efficient organisation of the defence industrial base will produce the politically desired results. By reserving the

---

19 The often renamed, presently labelled Euro-fighter provides an illustration of the old military design culture. This fighter aircraft will be equipped with a mechanical inertial device while civilian aircraft of the same generation have fibre optical inertial devices. A general trend prevails to restructure the military design culture towards growing integration of non-military technology these examples of the Euro-fighter notwithstanding.

most valuable economic resources for a closed defence industry the aspired results will not be achieved, because the implied technological self-sufficiency will produce only second rate material. The United States has even begun to accept the reality of the global economic competition speeding the rate of civilian innovation which the defence industries must absorb and integrate into their designs.

Russia will therefore only be able to maintain its defence industrial base at efficient levels if, and only if this sector is placed within the context of an industrial structure which moves rapidly towards full integration into the global industrial network. Russia's military technology will only have a competitive future, access to technological innovation and the necessary economic resources if it follows this path. The price of past excellence (the cost of which was never correctly established) was too high. A democratic government cannot afford to dedicate similar levels of resources to the military sector and military R & D.

Enterprises are too large and much too vertically integrated, R & D in the military sector is isolated and separated from productive application; management lacks market orientation. These attributes of the present industrial structures have to undergo significant changes if the transformation of the Russian economy is to succeed.

The last three years have clearly demonstrated that the management in charge of the large industrial conglomerates within the VPK behave primarily as political players and only in second place as business managers[20]. Future industrial structures are therefore more likely to evolve in competition with the old conglomerates instead of under their guidance; their role must at least be challenged from outside. If this assessment is correct, supporting conversion requires reinforcing of decentralizing tendencies[21] by lending support to the creation of new, smaller, industrial units.

---

20 Recent interviews with managers in the Northwestern region carried out by Ksenia Gonchar and Yevgeny Kuznetsov revealed an inclination of some managers to shed large parts of the workforce in order to concentrate on "lean production" largely for export, as soon as the legal situation will permit a "management buy-out privatisation".

21 The Kirovsky Zavod - 50,000 plus workers - in Saint Petersburg has been subdivided into dozens of small entities in a desperate attempt to save viable parts of this giant enterprise.

Existing programmes: Several international programmes in support of conversion and transformation within the Russian economy are being conducted in Saint Petersburg. Business seminars with simultaneous translation are most frequently offered; several countries offer a limited number of traineeships in industry and service sectors which require extensive and costly language preparation, - it may be considered an additional asset - the government of Baden-Würtemberg sponsors a business training centre which offers a variety of courses, some of which culminate in a three months practical training in Germany; the Danish government offers business classes including a period of practical training in Denmark.

Some of the more costly programmes contribute to creating a small group of persons whose qualification is eagerly sought by foreign companies who seek entry into the Russian market. These companies can afford to pay the highest salaries. Even though the candidates for these courses were chosen or mandated by large Russian enterprises, they are not likely to return to their former position. None of the shorter seminar-type programmes received a particularly positive rating in the interviews carried out in September 1993. The head of the conversion department in the City Council pointed out that the training courses were generally of little practical value, while the practical training abroad would lead regularly to business contacts with a potential for expansion.

To the knowledge of this author a programme has not yet reached the upper echelons of the large military enterprises[22]. This is not likely to change, not least for political reasons.

Proposal: Under the prevailing circumstances support of conversion should concentrate on creating alternative structures of production and services which would help to wean qualified personnel from the military-industrial complex or, positively, to allow these enterprises to shed parts of their workforce. Preparing a large number of persons with different levels of qualification to work in market-oriented private enterprises would make an important contribution to the transformation of the economy and help to

---

22   The inherited Soviet business culture and the high ranking of most of the enterprises in the old system prepared for defensive posture of the management. In conjunction with a systematic overestimating of the market value of their technological assets the enterprises failed to develop a realistic approach towards international cooperation and the necessary steps of restructuring in order to enter new civilian markets as subcontractors and suppliers of high-tech components.

convert or dissolve large sectors of the former military-industrial complex[23]. How to acquaint large numbers of otherwise well-trained and qualified people with the specific knowledge needed to operate successfully in the context of the evolving market economy in Russia constitutes the real challenge for a conversion strategy.

Training programmes should meet the following conditions as demand is, by definition, quasi-unlimited. Programmes should be cheap and should not require the command of a foreign language, they should take the specific conditions of transition in Russia into account, they should involve as far as possible Russian instructors (at least after the initial phase) and they should permit imitation.

In order to achieve these ends, it is proposed to engage in the creation of a small number of small private model enterprises whose purpose is to demonstrate the viability of such ventures and to facilitate on-the-job training of large numbers of trainees. These private model businesses should include typical activities in the service sector which the traditional Soviet enterprise will abandon: kindergartens, health clubs, shops, etc.

This approach builds on the positive experiences that many joint ventures have reported[24]. It is general practice to post a foreign specialist and a Russian counterpart in certain positions with the Russian counterpart taking over after a relatively short period. Although joint ventures reserve this pattern for higher positions and although they often purchase other required qualifications in the labour market by paying generally higher salaries, the proposed model enterprises are designed to serve as training centres for all levels of employment.

Manufacturing sector: It should be possible to find a workshop in one of the large military conglomerates which could be hived off to form a privatised model company with approx. 40 to 100 employees. After a necessary

---

23 Two recent empirical surveys commissioned by the World Bank show encouraging growth of the private sector in manufacturing as well as in the service sectors of Saint Petersburg. See: Webster, Leila and Joshua Charap, A Survey of Private Manufacturers in St. Petersburg Working Paper (Washington D.C.: The World Bank, May 1993); N.N., Private Service Firms in St. Petersburg: Findings of a Survey, Helsinki 1993.

24 For a recent report (on Otis's joint ventures) see: John Baxter, Russian Ups and Downs, Financial Times November 18, 1993, pg. 12.

headstart and a possible infusion of capital for modernisation of some equipment, the factory should operate in the real world of Russia's economy. All market-related positions should be staffed with an expert-instructor and a trainee; the trainees should remain in their positions for a period according to the specific requirements of learning on-the-job.

The transparency of all operations including critical situations would be important. In order to disseminate the experiences gained throughout the operation of the model plant, the whole experiment should be accompanied by a detailed coverage in an educational TV-series - possibly a mixture of reporting on the model enterprise and systematic lecturing of bookkeeping, business plans, etc.

Service sector: Following the same logic, a number of private establishments should be formed in the service sector. Training on the spot, and analytical TV-coverage of the activities and systematic lecturing on bookkeeping, marketing, legal counselling and others business processes would create a condition for large scale dissemination of knowledge and, hopefully, imitation. The selection of model services should cover the social activities formerly supplied by the large enterprises as was mentioned above.

What distinguishes the proposed approach from the hitherto practised schemes is the fact that the training reflects the day-to-day experience within the transition economy of Russia rather than the textbook advice developed in the context of established market economies. The model businesses should encourage imitation and could be tremendously facilitated because they operate in the same type of environment. It is also expected that the operation of model enterprises with full transparency would provide an important feedback for public institutions involved in regulating the market.

A foundation involving a number of Russian and foreign sponsors appears to be the adequate institutional framework for the creation of model enterprises at different levels and in different sectors of the economy.

This sketch intends to start a discussion on this novel approach[25] to support transition and indirectly conversion (which for the time being seems to be blocked for a number of diverse reasons).

---

25 Some preliminary cost calculations were made which can be supplied at a later stage of deliberation.

About the authors

Amosenok, Ella	Institute of Industrial Organisation, Novosibirsk
Bazhanov, Victor	Institute of Industrial Organisation, Novosibirsk
Gonchar, Ksenia	IMEMO, Moscow
Granqvist, Mathias	University Lund
Lock, Peter	Hamburg
Musienko, Igor	Stratecon, Novosibirsk
Opitz, Petra	University Oldenburg
Östh, Claes	University Lund
Pfaffenberger, Wolfgang	University Oldenburg
Salamon, Martin	OECD, Paris
Tomchin, Grigori	City administration St. Petersburg
Wedlin, Per	Swenesco, Stockholm
Whitman, Ian	OECD, Paris